OXFORD MEDICAL PUBLICATIONS

Cardiovascular homeostasis

Cardiovascular homeostasis

intrarenal and extrarenal mechanisms

SHEILA M. GARDINER
Research Fellow
and
TERENCE BENNETT
Reader in Physiology
Department of Physiology and Pharmacology,
University of Nottingham

Oxford
OXFORD UNIVERSITY PRESS
New York · Toronto
1981

Oxford University Press, Walton Street, Oxford OX2 6DP
London Glasgow New York Toronto
Delhi Bombay Calcutta Madras Karachi
Kuala Lumpur Singapore Hong Kong Tokyo
Nairobi Dar es Salaam Cape Town
Melbourne Wellington

and associate companies in
Beirut Berlin Ibadan Mexico City

© Sheila M. Gardiner and Terence Bennett, 1981

All rights reserved. No part of this publication may be reproduced, stored in a retrieval system, or transmitted, in any form or by any means, electronic, mechanical, photocopying, recording, or otherwise, without the prior permission of Oxford University Press

This book is sold subject to the condition that it shall not, by way of trade or otherwise, be lent, re-sold, hired or otherwise circulated without the publisher's prior consent in any form of binding or cover other than that in which it is published and without a similar condition including this condition being imposed on the subsequent purchaser

British Library Cataloguing in Publication Data

Gardiner, Sheila M
Cardiovascular homeostasis. – (Oxford medical publications).
1. Homeostasis
2. Cardiovascular system
I. Title   II. Bennett, Terence
612'.1   QP110.H6   80-41813
ISBN 0-19-261178-X

Set by Hope Services, Abingdon
Printed in Great Britain
at the University Press, Oxford
by Eric Buckley
Printer to the University

Acknowledgements

We wish to acknowledge the co-operation of the following publishers, and official bodies similarly concerned, in giving permission to reproduce copyright illustrations:

Acta Physiologica Scandinavica; American Association for the Advancement of Science (*Science*); American Heart Association (*Circulation Research*); American Physiological Society (*American Journal of Physiology; Disturbances of body fluid osmolality*); *British Medical Bulletin*; Elsevier/North-Holland (*Brain Research*); Federation of the American Society for Experimental Biology (*Federation Proceedings*); Grune & Stratton Inc. (*Metabolism*); S. Karger AG (*Neuroendocrinology*); J. B. Lippincott Co. (*Endocrinology*); Little, Brown & Co. (*Physiology,* fourth edition); Macmillan Journals Ltd. (*Nature*); Mayo Foundation (*Mayo Clinic Proceedings*); Physiological Society (*Journal of Physiology*); The Rockefeller University Press (*Journal of Clinical Investigation*); W. B. Saunders Co. Ltd. (*The kidney,* Volume 1); Springer Verlag (*Physiology of the kidney and of water balance; control mechanisms in drinking*); Year Book Medical Publishers (*Advances in nephrology*).

Full details of the source of illustrations are given in the references or Figure captions.

Contents

1. MICROCIRCULATION ... 1
   1.1. Capillary filtration coefficient ... 2
   1.2. Capillary hydrostatic pressure ... 3
      1.2.1. Myogenic mechanisms ... 4
      1.2.2. Metabolic influences ... 5
      1.2.3. Pre- and post-capillary resistances ... 6
      1.2.4. Sympathetic nervous mechanisms ... 7
   1.3. Interstitial fluid hydrostatic pressure ... 9
   1.4. Plasma and interstitial fluid colloid osmotic pressure ... 10
2. CARDIOVASCULAR REFLEXES ... 17
   2.1. Arterial baroreceptors ... 17
      2.1.1 Baroreceptor characteristics ... 17
      2.1.2. Central connections ... 20
      2.1.3. Baroreceptor reflexes ... 29
   2.2. Cardiopulmonary receptors ... 37
      2.2.1. Atrial receptors with vagal afferents ... 37
      2.2.2. Ventricular receptors with vagal afferents ... 41
      2.2.3. Cardiopulmonary receptors with sympathetic afferents ... 42
      2.2.4. Pulmonary receptors ... 43
   2.3. Chemoreceptors ... 44
3. REGULATION OF GLOMERULAR FILTRATION RATE AND RENAL BLOOD-FLOW ... 52
   3.1. Myogenic mechanisms ... 53
   3.2. Hormonal control ... 54
4. PROXIMAL TUBULAR REABSORPTION AND GLOMERULOTUBULAR BALANCE ... 63
   4.1. Sodium reabsorption ... 63
   4.2. Glomerulo-tubular balance ... 70
      4.2.1. Peritubular environment ... 71
      4.2.2. Intratubular environment ... 72
      4.2.3. Hormonal effects ... 72
      4.2.4. Sympathetic nervous effects ... 74

Contents

5. DISTAL REABSORPTION	80
5.1. Loop of Henle	80
5.2. Distal tubule	81
5.2.1. Action of aldosterone	81
5.2.2. Control of aldosterone secretion	84
6. RENIN RELEASE	92
6.1. Factors affecting renin release	92
6.1.1. Baroreceptor hypothesis	92
6.1.2. Macula densa hypothesis	94
6.1.3. Plasma sodium concentration	96
6.1.4. Plasma potassium concentration	97
6.1.5. Osmolality	97
6.1.6. Sympathetic nerve activity and circulating catecholamines	98
6.1.7. Hormones	99
6.2. Mechanism of renin release	100
7. ANGIOTENSIN	105
7.1. Peripheral effects of blood-pressure	105
7.2. Central effects on blood-pressure	108
8. ANTIDIURETIC HORMONE	114
8.1. Biosynthesis of antidiuretic hormone	114
8.2. Release of antidiuretic hormone	115
8.2.1. Osmoreceptors and sodium sensors	115
8.2.2. Volume receptors	119
8.2.3. Renin-angiotensin system	122
8.3. Actions of antidiuretic hormone	123
9. URINE CONCENTRATION AND DILUTION	131
9.1. Tubular characteristics	131
9.2. Renal countercurrent mechanisms	132
9.3. Countercurrent exchange	135
10. THIRST	139
10.1. Cellular dehydration	139
10.1.1. Central pathways	139
10.2. Extracellular fluid depletion	140
10.2.1. Renin-angiotensin system	142
10.2.2. Central pathways	144
Index	153

1. Microcirculation

The microcirculation provides a large surface area for the exchange of water and solutes between the interstitium and the vascular system. Anatomically the microcirculation consists of terminal arterioles that divide into smaller metarterioles which eventually give rise to true capillaries. At the distal end of the capillary bed, pairs of capillaries join to form post-capillary venules which merge into venules. The point of origin of the true capillaries may be guarded by a ring of smooth muscle – the pre-capillary sphincter – although in some vascular beds the metarterioles serve this function. The pre-capillary sphincter (or metarteriole) is the most distal point at which resistance changes can occur (Mellander and Johansson 1968), since true capillaries contain no smooth muscle.

The two principal aims of the control mechanisms which govern transcapillary fluid exchange are divergent. In some capillary beds the regulatory mechanisms are designed to *promote* fluid exchange and thereby control effective plasma volume, whereas in others the homeostatic mechanisms operate to *prevent* transcapillary fluid exchange and thereby hinder untoward fluid fluxes and hence oedema formation (Mellander 1978).

Starling's original concept of fluid movement being governed by the interaction of forces acting across the capillary wall (Starling 1896) is well established. These forces can be represented by the equation

$$F = K\left[(P_c - P_i) - (\pi_p - \pi_i)\right],$$

where F is the rate of fluid flow into or out of the capillary, K the filtration coefficient, P_c the capillary hydrostatic pressure, P_i the interstitial fluid hydrostatic pressure, π_p the plasma colloid osmotic pressure, and π_i the interstitial fluid colloid osmotic pressure.

Filtration or reabsorption occurs when the algebraic sum of the equation is positive or negative respectively. Starling (1896) originally suggested that filtration occurred primarily at the arterial end of the capillary and that a counterbalancing reabsorptive process occurred at the venular end, with zero net flux at the mid-capillary level. However, more recent evidence has shown that this may not be the case. For example, in the mesenteric capillary bed, filtration occurs along the entire length of the capillary and may extend into the venule (Zweifach

1971; Fraser, Smaje, and Verrinder 1978); under those conditions lymphatic drainage accounts for most of the removal of fluid from the interstitium (Fraser *et al.* 1978).

1.1. CAPILLARY FILTRATION COEFFICIENT

The capillary filtration coefficient is expressed in units of volume filtered per unit time, area, and pressure change, and hence reflects the capillary permeability and the surface area available for exchange. If single capillaries are occluded at different points along their lengths and exposed to different pressures, the relationship between fluid movement across the capillary (volume per unit time and area), and the measured capillary hydrostatic pressure is a straight line with a slope equal to the filtration coefficient (Landis 1927). A modification of this technique has since been applied to a number of different capillary beds (Zweifach and Intaglietta 1968; Smaje, Zweifach, and Intaglietta 1970; Michel, Mason, Curry, and Tooke 1974; Fraser *et al.* 1978: Table 1.1). The results presented in Table 1.1 show that the magnitude of the capillary filtration coefficient varies between, and even within,

Table 1.1 *Capillary filtration coefficients*

Tissue	Coefficient ($\mu m^3\ \mu m^{-2}\ s^{-1}\ N^{-1}\ m^2$)
Single capillaries	
Frog mesentery	$4-16 \times 10^{-5}$
Rabbit omentum:	
arterial end	$2-8 \times 10^{-5}$
venous end	$16-25 \times 10^{-5}$
Rat cremaster muscle	1×10^{-5}
Cat mesentery	
capillaries	$10-30 \times 10^{-5}$
collecting venules	$30-50 \times 10^{-5}$
Capillary beds	
Human forearm	1×10^{-6}
Dog hind-limb	2.5×10^{-6}
Cat hind-limb	$3.5-5.2 \times 10^{-6}$
Rabbit heart	8.6×10^{-6}
Dog lung	0.22×10^{-6}

Note that the data for capillary beds have been calculated assuming figures for the capillary surface area per unit weight of tissue. (With permission Caro *et al.* (1978).)

capillaries: it is greater at the venular end than at the arterial end, and generally lower in skeletal muscle than in the mesentery and omentum (Smaje et al. 1970; Fraser et al. 1978).

1.2. CAPILLARY HYDROSTATIC PRESSURE

Capillary hydrostatic pressure is the major determinant of transcapillary fluid movement since it shows the greatest fluctuations. Given that

$$\text{Flow} = \frac{\text{Pressure difference}}{\text{Resistance}},$$

then, in a capillary bed,

$$\text{inflow} = \frac{P_a - P_c}{r_a} \quad \text{and outflow} = \frac{P_c - P_v}{r_v},$$

where P_a is the arterial inflow pressure, P_c the capillary hydrostatic pressure, P_v the venous outflow pressure, r_a the pre-capillary resistance, and r_v the post-capillary resistance.

Under conditions of no net filtration or reabsorption,

$$\text{inflow} = \text{outflow}$$

Hence

$$\frac{P_a - P_c}{r_a} = \frac{P_c - P_v}{r_v}$$

$$(P_a - P_c)r_v = r_a(P_c - P_v)$$

$$P_a r_v - P_c r_v = P_c r_a - P_v r_a$$

$$P_c r_a + P_c r_v = P_a r_v + P_v r_a.$$

Therefore

$$P_c = \frac{P_a(r_v/r_a) + P_v}{1 + (r_v/r_a)}$$

(Pappenheimer and Soto-Rivera 1948).

Thus capillary hydrostatic pressure depends on arterial pressure, venous pressure, and the ratio between pre- and post-capillary resistances. In some capillary beds, the hydrostatic pressure is relatively independent of systemic arterial pressure (Zweifach 1971, 1974), and it is likely that this is due to the factors which influence blood-flow to those vascular beds. These factors are considered below.

1.2.1. Myogenic mechanisms

The phenomenon of auto-regulation of blood-flow is well documented (Folkow 1964; Johnson 1964; Robard 1971). Jones and Berne (1964) showed that changing the pressure in a perfused, denervated vascular bed in the hind-limb of the dog caused an immediate change in blood-flow but within 30–60 seconds after the manoeuvre, blood-flow was restored to control levels (Fig. 1.1). To maintain flow constant in the

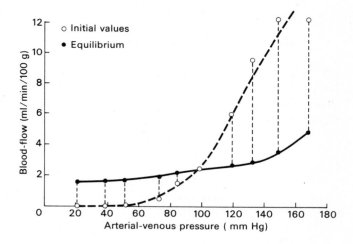

Fig. 1.1. Pressure–flow relationships in resting acutely denervated skeletal muscle. All pressure changes were initiated from a control perfusion pressure of 110 mm Hg. Open circles represent the maximum or minimum flow change immediately following the pressure change whereas closed circles represent the stable flow values reached one to three minutes following the sustained pressure change. (With permission, Jones and Berne (1964).)

face of increased pressure, resistance must necessarily increase since flow is equal to the pressure difference divided by the resistance. Bayliss (1902) was the first to observe that increasing the perfusion pressure of an isolated carotid artery preparation caused a powerful contraction of the vessel (which was seen to 'writhe like a worm'); when the perfusion pressure was lowered the vessel relaxed. Bayliss suggested that the vascular smooth muscle was stimulated by being stretched (Bayliss 1902). However, if this were the case, an increase in perfusion pressure would stretch the vessel and trigger a contraction, but the stimulus would be removed when the vessel diameter was back to normal. It is therefore more likely that a change in transmural tension is the signal perceived by the vascular smooth muscle. The law of Laplace states that wall tension is proportional to the product of the distending pressure and the vessel radius. An increase in perfusion pressure would passively distend the vessel and thereby cause a large increase in wall tension, leading to a contraction. If the vessel radius only returned to the control level, then the product of the pressure (increased) and radius (normal) would still be high and the stimulus for contraction would still be present. Only when the vessel radius was less than control (the requisite for flow regulation), could the product of pressure (increased) and radius (decreased) be normal, and hence the stimulus for contraction removed.

The above-mentioned process has been termed 'myogenic auto-regulation' since it occurs in isolated preparations and in denervated vascular beds, and must therefore be a manifestation of some inherent property of the vascular smooth muscle. The mechanisms involved are not clearly defined; Folkow (1964) suggested that the smooth muscle cells of the resistance vessels constituted a population of 'pace-makers' which increased their frequency of discharge when stretched. Alternatively, Johnson (1974, 1977) suggested that there was a sensor element in series with a contractile element in the vascular smooth muscle. An increase in pressure would increase the wall tension and stretch the sensor element, and this would stimulate the contractile element; only when the vessel diameter was less than before would the stimulus be abolished (Johnson 1974, 1977). Recent evidence (Mellander 1978) has shown that there is both a static and a dynamic component to the myogenic response to increased transmural pressure. Thus, in addition to the level of distension, the rate of change of pressure may determine the magnitude of the contraction (Mellander 1978).

1.2.2. Metabolic influences

Several workers have suggested that auto-regulation of blood-flow is brought about by the accumulation of vasodilator metabolites (e.g. Haddy and Scott 1971; Jonsson 1971; Skinner and Costin 1971;

Granger, Goodman, and Cook 1975) and it has been claimed that their vasodilator effects may be due to changes in osmolality which they cause (Jonsson 1971).

The metabolic theory of auto-regulation does not explain the responses observed by Bayliss (1902) using a perfused, isolated, carotid artery preparation. Nevertheless it may be that vasodilator metabolites interact with the myogenic mechanism to control blood-flow. Thus, an increase in resistance due to a myogenic contraction could provoke a buildup of metabolic products, thereby causing a certain degree of vasodilatation. In this way metabolic responses might act as a 'brake' on the myogenic control mechanisms (Folkow 1964; Johnson 1967; Johnson and Henrich 1975).

Elevation of venous pressure causes metabolic and myogenic mechanisms to come into opposition. The resultant rise in intravascular pressure constitutes a stimulus for myogenic constriction, while at the same time capillary blood-flow falls, giving rise to conditions favouring metabolic vasodilatation. Under these conditions Johnson (1967) observed a vasoconstrictor response, indicating the dominance of myogenic mechanisms. However, if a vascular bed is perfused at very low rates there is a fall in resistance in response to increased venous pressure (Shepherd 1977), suggesting that the myogenic response is abolished.

1.2.3. Pre- and post-capillary resistances

If the myogenic response to increased perfusion pressure occurred equally in pre- and post-capillary vessels, it would be possible to have a situation where flow was regulated but capillary hydrostatic pressure was elevated. However, Järhult and Mellander (1974) showed that, in denervated lower leg muscles of the cat, myogenic constriction occurred predominantly in the pre-capillary vessels such that the pre- to post-capillary resistance ratio rose with increasing perfusion pressure. They demonstrated that regional blood-pressure changes between 170 and 30 mm Hg were accompanied by only small changes in capillary hydrostatic pressure; in the hypotensive range (20–95 mm Hg) the change in capillary pressure was ± 0.5 mm Hg, and in the hypertensive range (95–180 mm Hg) the maximal change in capillary pressure was 2 mm Hg (Järhult and Mellander 1974). Likewise Zweifach (1971, 1973, 1974) showed that when mean systemic arterial blood-pressure fell from 90 to 60 mm Hg the capillary pressure in the omentum changed by only

2 mm Hg. These workers argued that the auto-regulation of capillary pressure provided a defence against changes in arterial blood-pressure causing untoward fluid fluxes across the capillary endothelium (Järhult and Mellander 1974; Zweifach 1974). However, Gore (1974) and Gore and Bohlen (1975) questioned the benefit of capillary pressure autoregulation, since the appropriate response to a fall in arterial pressure, due to fluid loss, would be increased transcapillary reabsorption. Gore (1974) measured hydrostatic pressure in mesenteric capillaries during changes in systemic arterial blood-pressure. In some vessels (five out of eight) there was no change in capillary hydrostatic pressure over a wide range of systemic arterial pressures, but in the remaining vessels there was a linear relationship between mean systemic arterial blood-pressure and capillary hydrostatic pressure. Gore (1974) suggested that capillary hydrostatic pressure auto-regulation was merely a reflection of regulation of blood-flow and was entirely attributable to the peculiar anatomical organization of the mesenteric microcirculation. This claim is not consistent with the observations of Järhult and Mellander (1974) on auto-regulation of capillary hydrostatic pressure in skeletal muscle (see above).

Subsequent experiments (Gore and Bohlen 1975; Bohlen and Gore 1977) showed a direct correlation between systemic arterial bloodpressure and intestinal capillary hydrostatic pressure. However, the validity of these findings is questionable since isoprenaline was used to suppress intestinal motility and it has been shown that isoprenaline effectively abolishes capillary hydrostatic pressure auto-regulation (Järhult and Mellander 1974).

1.2.4. Sympathetic nervous mechanisms

The majority of evidence favours the view that many vascular beds have the inherent ability to regulate capillary hydrostatic pressure. However, in addition, nervous and humoral mechanisms can interact with, and sometimes override, the local control processes such that transcapillary fluid fluxes may occur. The sympathetic nervous system is particularly important in this respect. Mellander (1960) first demonstrated that α-adrenoceptor stimulation increased the pre- to postcapillary resistance ratio, and thereby promoted net transcapillary fluid reabsorption. Later Öberg (1964) showed that afferent signalling from cardiac mechanoreceptors and arterial baroreceptors caused reflex vasomotor activity which had particularly marked effects on the metarterioles in skin and skeletal muscle; the resultant fall in pre- to post-capillary resistance ratio caused capillary hydrostatic pressure to fall, favouring fluid reabsorption into the vascular space. Sympathetic vasoconstrictor mechanisms predominantly affect the arterioles rather than the pre-capillary sphincters (Honig, Frierson, and Patterson 1970). However, there is a β-adrenoceptor-mediated dilator mechanism which affects the pre-capillary sphincters and small resistance vessels (Lundvall and Järhult 1976). Whereas the constrictor mechanism controls fluid reabsorption by altering the pre- to post-capillary resistance ratio, the dilator mechanism enhances fluid reabsorption by increasing the number of patent capillaries and hence the surface area available for exchange (Lundvall and Järhult 1976). Beta-adrenoceptor stimulation may also supplement the fall in capillary hydrostatic pressure brought about by the α-adrenoceptor-mediated constrictor mechanism by causing a relatively greater vasodilatation in post- than in pre-capillary vessels (Lundvall and Järhult 1976).

Lundvall and Hillman (1978) showed, using the lower leg muscles of the cat, that haemorrhage (15 per cent of the blood volume) for 10 minutes caused an 80 per cent increase in vascular resistance and also a pre-capillary vasodilatation which was manifest as a 35 per cent increase in filtration coefficient; under these conditions there was a reabsorption of extravascular fluid. The same procedure in the presence of local β-adrenoceptor blockade caused an even greater increase in vascular resistance (110 per cent) but pre-capillary sphincter tone and hence transcapillary fluid reabsorption were unaffected (Lundvall and Hillman 1978). During prolonged haemorrhage (longer than 120 minutes) there is evidence that the vascular responsiveness to constrictor stimuli wanes as vasodilator metabolites accumulate (Mellander and Lewis 1963). This occurs in pre-capillary vessels sooner than in post-capillary vessels, leading to a decreased pre- to post-capillary resistance ratio. The resultant rise in capillary hydrostatic pressure increases the fluid loss from the vascular space and leads, ultimately, to haemorrhagic shock and circulatory failure (Mellander and Lewis 1963).

Sympathetically-mediated constriction of veins may also play some part in the control of capillary hydrostatic pressure. However, there is some disagreement as to whether it is only in very severe stress states that active venoconstriction occurs (Gauer, Henry, and Behn 1970), or whether the degree of venoconstriction is linearly related to the extent of volume depletion (Drees and Rothe 1974).

The above-mentioned sympathetic mechanisms contribute to the restoration of effective plasma volume by a change in capillary hydrostatic pressure causing reabsorption of interstitial fluid into the vascular space. There is another adrenergically-mediated process which assists in restoring plasma volume by causing intracellular water to move into the extracellular space. During haemorrhage, hyperglycaemia develops due to hepatic glycogenolysis (Järhult 1975). The resultant extracellular hyperosmolality (up to 20 mOsm/kg) causes the withdrawal of intracellular fluid and hence extracellular fluid volume reexpansion (Järhult 1975).

1.3. INTERSTITIAL FLUID HYDROSTATIC PRESSURE

Transcapillary fluid exchange also depends on the interstitial fluid hydrostatic pressure which is influenced by the volume of fluid in the interstitium and the compliance of the organ in question. As yet, no entirely satisfactory method has been devised for accurately measuring the interstitial fluid hydrostatic pressure. Some workers (McMaster 1946; Kjellmer 1964) have used a micropuncture technique and obtained values between 0 and 5 mm Hg. However, micropuncture inevitably causes some tissue damage and distortion which probably leads to erroneous results. Another technique which has been used involves implanting a small, hollow, porous capsule, subcutaneously (Guyton 1963). The capsule is then left for three to four weeks, during which time its surface becomes covered with loose connective tissue and its centre filled with fluid, which is presumed to be in equilibrium with interstitial fluid. A fine needle is then inserted into the capsule and pressure is recorded. The values obtained from this technique are usually subatmospheric (-6 mm Hg; Guyton 1963). The objection to this method is that the connective tissue coating might constitute a semipermeable membrane around the capsule which could act as a barrier against the movement of interstitial protein. As a result, an effective osmotic gradient could develop which would tend to draw water out of the capsular space thereby creating a negative intracapsular pressure (Haljamäe, Linde, and Amundson 1974). In support of this argument, Stromberg and Wiederhielm (1970) showed that dilution of the interstitial protein with Tyrode's solution caused the hydrostatic pressure in the capsule to become positive. The most recent technique for measuring interstitial fluid hydrostatic pressure employs a saline-filled, cotton-wool wick system (Scholander, Hargens, and Miller 1968; Snashall, Lucas, Guz, and Floyer 1971; Clough and Smaje 1978). Long-stranded cotton wool is placed inside a polyethylene catheter with approximately 1 cm of wick exposed at one end and a pressure transducer fixed to the other. The whole system is filled with isotonic saline and the wick end of the cannula is implanted subcutaneously; the assumption is that fibres of the wick provide channels for hydrostatic communication between the interstitium and the saline-filled catheter. The values of interstitial fluid hydrostatic pressure obtained with this technique are also subatmospheric (-0.7 mm Hg, Clough and Smaje 1978). The objection to this technique pertains to the structure of the interstitium. It was earlier assumed that the interstitium was effectively a single phase, but there is now increasing evidence to show that the interstitium has two components — a 'sol' phase in which the proteins are contained, and a 'gel' phase made up of aggregated mucopolysaccharides which tend to imbibe water (Wiederhielm 1968; Laurent 1970). Since the proteins are effectively barred from the 'gel' phase the colloid osmotic pressure of the 'sol' phase is greater than that of plasma. Hence, when a catheter system filled with isotonic saline is placed in the interstitium, water will tend to be drawn out of the catheter and a negative pressure recorded (Caro, Pedley, Schroter, and Seed 1978). Some support for this assertion comes from the work of Clough and Smaje (1978) who showed that adding 5 per cent albumin to the saline in the catheter made the interstitial fluid pressure more positive; nevertheless the value was still subatmospheric (-0.5 mm Hg; Clough and Smaje 1978). Thus at the present time the general opinion is that interstitial fluid hydrostatic pressure is slightly subatmospheric.

1.4. PLASMA AND INTERSTITIAL FLUID COLLOID OSMOTIC PRESSURES

The plasma colloid osmotic pressure and the interstitial fluid colloid osmotic pressure are the two remaining forces which affect transcapillary fluid exchange. Circulating protein levels are chiefly determined by liver synthesis and degradation and by the movement of protein between the vascular space and the interstitium. The latter process (which necessarily affects interstitial fluid colloid osmotic pressure) depends on the permeability of the capillary endothelium to protein. Muscle capillaries are not very permeable to protein whereas intestinal and hepatic capillaries have a higher permeability to protein (Landis and Pappenheimer 1963; Friedman 1976). Protein movement from the interstitium into the vascular space is normally very limited although, in a situation where a substantial amount of fluid is being drawn into the vascular space, some protein may move with this fluid (Wasserman, Joseph, and Mayerson 1956; Skillman, Awwad, and Moore 1967; Reeve and Chen 1970). During haemorrhage, a fall in capillary hydrostatic pressure increases transcapillary fluid reabsorption, which might be expected to lower the plasma colloid osmotic pressure; such a situation would antagonize reabsorption. However, under these conditions some protein moves from the interstitium into the vascular space (see above), thereby attenuating the fall in colloid osmotic pressure. Furthermore, protein synthesis in the liver is triggered by a number of events during haemorrhage (Skillman *et al.* 1967; Hoffenberg 1970; Carey 1973).

In summary the forces acting across the capillary walls are, under resting conditions, stabilized by mechanisms inherent to the vascular bed; this maintains a constant capillary blood-flow and pressure and prevents transcapillary fluid fluxes. However, the latter may occur when nervous and/or humoral mechanisms are activated and they serve to restore effective plasma volume. Fig. 1.2 (next page) shows the effect on transcapillary fluid movement of arterial and venous pressures, resistances, and plasma colloid osmotic pressure.

REFERENCES

Bayliss, W. M. (1902). On the local reactions of the arterial wall to changes in internal pressure. *J. Physiol.* 28, 220-31.

Bohlen, H. G. and Gore, R. W. (1977). Comparison of microvascular pressures and diameters in innervated and denervated rat intestine. *Microvasc. Res.* 14, 251-64.

Carey, J. S. (1973). Physiological hemodilution: interrelationships between hemodynamics and blood volume after acute blood loss. *Ann. Surg.* 178, 87-94.

Caro, C. G., Pedley, T. J., Schroter, R. C., and Seed, W. A. (1978). The systemic circulation. In *The mechanics of the circulation*, Chapter 13, pp. 350-433. Oxford University Press.

Clough, G. and Smaje, L. H. (1978). Simultaneous measurement of pressure in the interstitium and the terminal lymphatics of the cat mesentery. *J. Physiol.* 283, 457-68.

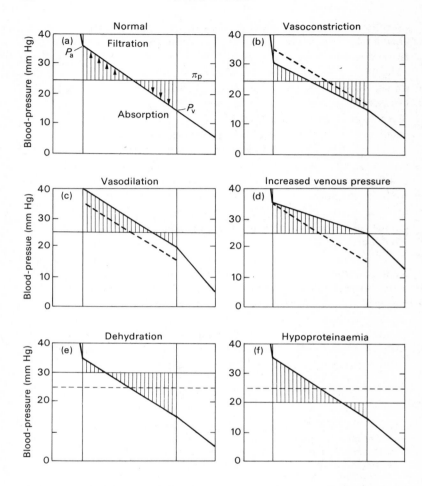

Fig. 1.2. Variations in capillary filtration and absorption produced by changes in arterial (P_a) and venous (P_v) pressures and resistances and plasma colloid osmotic pressure (π_p). Where there is a change in these variables, the broken line represents the normal values. Capillary filtration is increased with vasodilation (c), elevated venous pressure (d), and hypoproteinaemia (f); reabsorption is increased by vasoconstriction (b) and dehydration (e). (With permission, Friedman (1976).)

Drees, J. A. and Rothe, C. F. (1974). Reflex venoconstriction and capacity vessel pressure–volume relationships in dogs. *Circulation Res.* **34**, 360–73.

References

Folkow, B. (1964). Description of the myogenic hypothesis. *Circulation Res.* **15** (Suppl I), 279–87.
Fraser, P. A., Smaje, L. H., and Verrinder, A. (1978). Microvascular pressures and filtration coefficients in the cat mesentery. *J. Physiol.* **283**, 439–56.
Friedman, J. J. (1976). Microcirculation. In *Physiology* (ed. E. E. Selkurt), pp. 273–88. Little, Brown, & Co, Boston.
Gauer, O. H. Henry, J. P., and Behn, C. (1970). The regulation of extracellular fluid volume. *Ann. Rev. Physiol.* **32**, 547–95.
Gore, R. W. (1974). Pressures in cat mesenteric arterioles and capillaries during changes in systemic arterial blood pressure. *Circulation Res.* **34**, 581–91.
—— and Bohlen, H. G. (1975). Pressure regulation in the microcirculation. *Fed. Proc.* **34**, 2031–7.
Granger, H. J., Goodman, A. H., and Cook, B. H. (1975). Metabolic models of microcirculatory regulation. *Fed. Proc.* **34**, 2025–30.
Guyton, A. C. (1963). A concept of negative interstitial pressure based on pressures in implanted perforated capsules. *Circulation Res.* **12**, 399–414.
Haddy, F. J. and Scott, J. B. (1971). Bioassay and other evidence for participation of chemical factors in local regulation of blood flow. *Circulation Res.* **28–9** (Suppl. I), 86–92.
Haljamäe, H., Linde, A., and Amundson, B. (1974). Comparative analyses of capsular fluid and interstitial fluid. *Amer. J. Physiol.* **227**, 1199–205.
Hoffenberg, R. (1970). Control of albumin degradation *in vivo* and in the perfused liver. In *Plasma protein metabolism. Regulation of synthesis, distribution and degradation* (ed. M. A. Rothschild, and T. Waldmann), pp. 239–55. Academic Press, New York & London.
Honig, C. R., Frierson, J. L., and Patterson, J. L. (1970). Comparison of neural controls of resistance and capillary density in resting muscle. *Amer. J. Physiol.* **218**, 937–42.
Järhult, J. (1975). Osmolar concentration of the circulation in hemorrhagic hypotension. An experimental study in the cat. *Acta physiol. Scand. Suppl.* **423**.
—— and Mellander, S. (1974). Autoregulation of capillary hydrostatic pressure in skeletal muscle during regional arterial hypo- and hypertension. *Acta physiol. Scand.* **91**, 32–41.
Johnson, P. C. (1964). Review of previous studies and current theories of autoregulation. *Circulation Res.* **15** (Suppl. 1), 2–9.
—— (1967). Autoregulation of blood flow in the intestine. *Gastroenterology* **52**, 435–41.
—— (1974). The microcirculation, and local and humoral control of the circulation. In *Cardiovascular physiology* (ed. A. C. Guyton and C. E. Jones), MTP International Review of Science, Physiology Ser. 1, Vol. 1, pp. 163. University Park Press: Butterworths.

Johnson, P. C. (1977). The myogenic response and the microcirculation. *Microvasc. Res.* **13**, 1–18.
—— and Henrich, H. A. (1975). Metabolic and myogenic factors in local regulation of the microcirculation. *Fed. Proc.* **34**, 2020–4.
Jones, R. D. and Berne, R. M. (1964). Intrinsic regulation of skeletal muscle blood flow. *Circulation Res.* **14**, 126–38.
Jonsson, O. (1971). Extracellular osmolality and vascular smooth muscle activity. *Acta physiol. Scand. Suppl.* **359**.
Kjellmer, I. (1964). An indirect method for estimating tissue pressure with special reference to tissue pressure in muscle during exercise. *Acta physiol. Scand.* **62**, 31–40.
Landis, E. M. (1927). Micro-injection studies of capillary permeability. II. The relation between capillary pressure and the rate at which fluid passes through the walls of single capillaries. *Amer. J. Physiol.* **82**, 217–38.
—— and Pappenheimer, J. R. (1963). Exchange of substances through the capillary walls. In *Handbook of physiology*, Sect. 2, Vol. II (ed. W. F. Hamilton), pp. 961–1034. American Physiological Society, Washington DC.
Laurent, T. C. (1970). The structure and function of the intercellular polysaccharides in connective tissue. In *Capillary permeability. The transfer of molecules and ions between capillary blood and tissue. Proceedings of the Alfred Benzon Symposium II* (ed. C. Crone and N. A. Lassen), pp. 261–77. Scandinavian University Books; Munksgaard.
Lundvall, J. and Hillman, J. (1978). Fluid transfer from skeletal muscle to blood during hemorrhage. Importance of beta adrenergic vascular mechanisms. *Acta physiol. Scand.* **102**, 450–8.
—— and Järhult, J. (1976). Beta adrenergic dilator component of the sympathetic vascular response in skeletal muscle. Influence on the microcirculation and on transcapillary exchange. *Acta physiol. Scand.* **96**, 180–92.
McMaster, P. D. (1946). The pressure and interstitial resistance prevailing in the normal and oedematous skin of animals and man. *J. exp. Med.* **84**, 473–94.
Mellander, S. (1960). Comparative studies on the adrenergic neurohormonal control of resistance and capacitance blood vessels in the cat. *Acta physiol. Scand. Suppl.* **176**.
—— (1978). On the control of capillary fluid transfer by precapillary and postcapillary vascular adjustments. A brief review with special emphasis on myogenic mechanisms. *Microvasc. Res.* **15**, 319–30.
—— and Johansson, B. (1968). Control of resistance, exchange and capacitance functions in the peripheral circulation. *Pharmacol. Rev.* **20**, 117–96.
—— and Lewis, D. H. (1963). Effect of hemorrhagic shock on the reactivity of resistance and capacitance vessels and on capillary filtration transfer in cat skeletal muscle. *Circulation Res.* **13**, 105–18.

References

Michel, C. C., Mason, J. C., Curry, F. E., and Tooke, J. E. (1974). A development of the Landis technique for measuring the filtration coefficient of individual capillaries in the frog mesentery. *Quart. J. exp. Physiol.* **59**, 283-309.
Öberg, B. (1964). Effects of cardiovascular reflexes on net capillary fluid transfer. *Acta physiol. Scand. Suppl.* **229**.
Pappenheimer, J. R. and Soto-Rivera, A. (1948). Effective osmotic pressure of the plasma proteins and other quantities associated with the capillary circulation in the hindlimbs of cats and dogs. *Amer. J. Physiol.* **152**, 471-91.
Reeve, E. B. and Chen, A. Y. (1970). Regulation of interstitial albumin. In *Plasma protein metabolism. Regulation of synthesis, distribution and degradation* (ed. M. A. Rothschild and T. Waldmann), pp. 89-109. Academic Press, New York & London.
Robard, S. (1971). Capillary control of blood flow and fluid exchange. *Circulation Res.* **28-9** (Suppl. I), 51-8.
Scholander, P. F., Hargens, A. R., and Miller, S. L. (1968). Negative pressure in the interstitial fluid of animals. *Science* **161**, 321-8.
Shepherd, A. P. (1977). Myogenic responses of intestinal resistance and exchange vessels. *Amer. J. Physiol.* **233**, H547-H554.
Skillman, J. J., Awwad, H. K., and Moore, F. D. (1967). Plasma protein kinetics of the early transcapillary refill after hemorrhage in man. *Surg. Gynecol. Obstet.* **125**, 983-96.
Skinner, N. S., Jr. and Costin, J. C. (1971). Interactions between oxygen, potassium and osmolality in regulation of skeletal muscle blood flow. *Circulation Res.* **28** (Suppl. I), 73-85.
Smaje, L. H., Zweifach, B. W., and Intaglietta, M. (1970). Micropressure and capillary filtration coefficients in single vessels of the cremaster muscle of the rat. *Microvasc. Res.* **2**, 96-110.
Snashall, P. D., Lucas, J., Guz, A., and Floyer, M. A. (1971). Measurement of interstitial 'fluid' pressure by means of a cotton wick in man and animals. An analysis of the origin of the pressure. *Clin. Sci.* **41**, 35-53.
Starling, E. H. (1896). On the absorption of fluids from the connective tissue spaces. *J. Physiol.* **19**, 312-26.
Stromberg, D. D. and Wiederhielm, C. A. (1970). Effects of oncotic gradients and enzymes on negative pressures in implanted capsules. *Amer. J. Physiol.* **219**, 928-32.
Wasserman, K., Joseph, J. D., and Mayerson, H. S. (1956). Kinetics of vascular and extravascular protein exchange in unbled and bled dogs. *Amer. J. Physiol.* **184**, 175-82.
Wiederhielm, C. A. (1968). Dynamics of transcapillary fluid exchange. *J. gen. Physiol.* **52**, 29-63.
Zweifach, B. W. (1971). Local regulation of capillary pressure. *Circulation Res.* **28-9** (Suppl. I), 129-34.
—— (1973). Microcirculation. *Ann. Rev. Physiol.* **35**, 117-50.

Zweifach, B. W. (1974). Quantitative studies of microcirculatory structure and function: 1. Analysis of pressure distribution in the terminal vascular bed in cat mesentery. *Circulation Res.* **34**, 843-57.
—— and Intaglietta, M. (1968). Mechanics of fluid movement across single capillaries in the rabbit. *Microvasc. Res.* **1**, 83-101.

2. Cardiovascular reflexes

There are sensory receptors associated with the cardiovascular system whose afferent signalling can reflexly modulate vascular resistance and the rate and force of beating of the heart and thereby help to maintain systemic arterial blood-pressure and, in certain cases, tissue perfusion. Inputs from different sites may interact centrally to enhance each others effects (see Abboud 1979), although there are instances in which activation of certain receptors leads to a suppression of the expected reflex responses (see later).

This chapter outlines the principal types of cardiovascular receptors and the reflex effects which occur as a result of their stimulation; the central connections involved in these cardiovascular reflexes are also considered.

2.1. ARTERIAL BARORECEPTORS

2.1.1. Baroreceptor characteristics

Afferent nerve endings mainly associated with the carotid sinus and aortic arch are collectively referred to as arterial baroreceptors. The sensory endings are not sensitive to intraarterial pressure changes as such, since if an increase in intrasinus pressure is prevented from causing distension of the vessel, then baroreceptor reflexes are not activated (Hauss, Kreuziger, and Asteroth 1949). Angell-James (1971) has produced clear evidence that aortic transmural pressure is the stimulus for aortic arch baroreceptor activation (Fig. 2.1). This has important consequences in terms of the ability of arterial baroreceptors to contribute to cardiovascular homeostasis, since disease processes affecting carotid sinus or aortic arch distensibility will influence the sensitivity of baroreceptors to changes in intravascular pressures.

Both the carotid sinus and aortic arch are, normally, relatively distensible over a range corresponding to low or moderate intravascular pressures (up to 120–40 mm Hg); above this level the vessels are much less easily distended. These mechanical characteristics of the vessels are largely responsible for the fact that baroreceptor afferents show a plateau in their discharge rate of action potentials even though intravascular pressure may not be maximal (see Kirchheim 1976). The observation that changes in pulse pressure and pulse frequency might

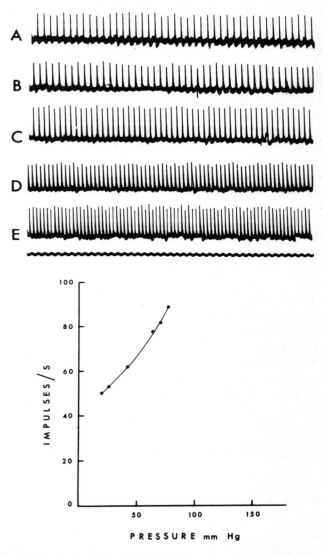

Fig. 2.1. The effects of reducing the extramural pressure on the impulse frequency of a single fibre from the left aortic nerve of a rabbit (A–E). Intra-aortic perfusion pressure was maintained constant at 20 mm Hg. The graph shows the relationship between impulse frequency and transmural pressure. (With permission, Angell-James (1971).)

influence baroreceptor signalling (Schmidt, Kumada, and Sagawa 1972) could be attributable to the dynamic characteristics of the baroreceptor afferents themselves, rather than to the structure of the vessel walls.

Several investigators (see Kirchheim 1976) have shown that application of α-adrenoceptor agonists onto the carotid sinus causes reflex cardiovascular effects, but attempts to reproduce such effects by stimulating the efferent sympathetic supply to the sinus have produced equivocal results. However, Keith, Kidd, Malpus, and Penna (1974) re-investigated the problem in chloralose-anaesthetized dogs. They recorded afferent discharges from baroreceptors associated with the right subclavian–brachiocephalic junction, and showed that the discharge rate was reduced by stimulation of the right ansa subclavia; this effect was abolished by phenoxybenzamine. Keith *et al.* (1974) suggested that the baroreceptor inhibition might have been due to neuronally-released noradrenalin affecting the wall of the carotid sinus or afferent nerve terminals directly.

More recently Peveler, Bergel, Gupta, Sleight, and Worley (1980) carried out experiments on vascularly isolated carotid sinuses in anaesthetized dogs. They measured the carotid sinus diameter and baroreceptor discharge and observed the effects of stimulating the vagosympathetic nerve trunk. Peveler *et al.* (1980) found that, at any constant sinus pressure, the diameter decreased in response to efferent nerve stimulation. Such nerve stimulation reduced the adapted discharge rate of baroreceptor fibres when carotid sinus pressure was low, but increased it when carotid sinus pressure was high. There have been no systematic studies on the involvement of sympathetic efferents to the carotid sinus in cardiovascular reflexes mediated via arterial baroreceptors, so the functional significance of observations such as those of Keith *et al.* (1974) and Peveler *et al.* (1980) remains to be elucidated. However, it is clear that pathways exist whereby the central nervous system could influence the sensitivity of arterial baroreceptors.

Recently Kunze and Brown (1978) described another variable which has direct effects on the sensitivity of arterial baroreceptors to changes in carotid sinus pressure. They showed that changes in the sodium concentration of the extracellular fluid bathing baroreceptor endings affected cardiovascular and renal function. A five per cent reduction of extracellular sodium concentration evoked a significant reflex increase in systemic arterial blood-pressure and urine flow, with no change in sodium excretion (Fig. 2.2). It is possible that this sodium sensitivity of

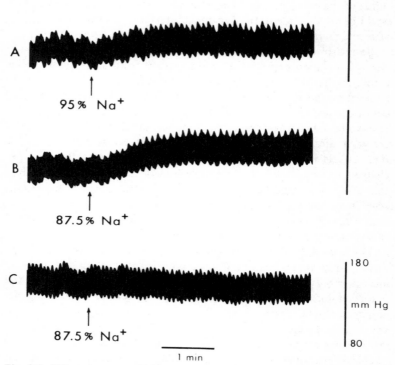

Fig. 2.2. Effects on systemic blood-pressure of changing carotid sinus perfusate sodium concentration, $[Na^+]_o$, from 145 to 138 mM (A) and from 145 to 127 mM (B). Twenty minutes before trace C was taken, the carotid sinus nerve was cut. Then, 127 mM $[Na^+]_o$ test perfusate was re-introduced. No change in pressure occurred. (With permission, Kunze and Brown (1978).)

arterial baroreceptors is important in the long-term control of blood-pressure, rather than in acute homeostasis.

2.1.2. Central connections

The cell bodies of the arterial baroreceptors are located in the nodose ganglion. Their afferent fibres enter the medulla in the IXth and Xth (glossopharyngeal and vagus) cranial nerves at a level close to the obex, and relay in the region of the nucleus of the tractus solitarius (NTS: Korner 1971). Early histological studies showed that the afferents terminated solely within the NTS (Cottle 1964) whereas later results, based on recordings of orthodromic action potentials, suggested that there was a wider distribution of terminals (Miura and Reis 1972). However, Chiba and Doba (1975) showed that either removal of the nodose ganglion or transection of the IXth and Xth cranial nerves caused degeneration of about 20 per cent of the total population of axon terminals in the NTS, but no degeneration in any other cell group. Furthermore, the more recent technique of measuring antidromic activity in carotid sinus and aortic nerves during stimulation of various sites in the medulla (Jordan and Spyer 1977) has provided evidence that the arterial baroreceptor fibres terminate solely within the region of the NTS (Garcia, Jordan, and Spyer 1979). The central pathways involved in the reflex responses to cardiovascular afferent stimulation are polysynaptic, but the primary synapses reside in the NTS and it is from there that secondary pathways arise. Connecting neurons pass from the NTS into the medullary reticular formation.

Experimental work during the last century (for review see Folkow and Neil 1971) showed that serial transections through the medulla could alter blood-pressure, the change depending on the level of section. This work led to the delineation of pressor and depressor areas in the medulla–pons region of the brain; pressor areas were found to extend more rostrally and to be situated more laterally than depressor areas. Electrical stimulation of the depressor areas increases vagal efferent tone and concurrently inhibits peripheral sympathetic nerve activity, whilst electrical stimulation of the pressor areas increases efferent sympathetic tone and evokes adrenal catecholamine release. Firing of the neurons in the depressor areas depends entirely upon afferent activity, whereas the neurons in the pressor regions are tonically active, but are influenced by afferent inputs from cardiovascular receptors. The cell bodies of the cardiac vagal efferent neurons (the cardio-inhibitory centre) are also in the medullary reticular formation, principally in the nucleus ambiguus but also in the dorsal motor nucleus in some species (Gunn, Sevelius, Puiggari, and Myers 1968; Korner 1971; Thomas and Calaresu 1974; McAllen and Spyer 1976; Manning 1977). The vagal cardio-inhibitory centre, like the depressor area, is not tonically active. The cell bodies in the nucleus ambiguus receive an excitatory input from the arterial baroreceptors and section of the aortic and carotid sinus nerves virtually abolishes vagal tone. There is neuroanatomical evidence for a direct connection between the NTS and the nucleus ambiguus (Cottle and Calaresu 1975; Loewy and Burton

1978) although, on the basis of measurements of the latency between stimulation of carotid sinus baroreceptors and activation of cardiac vagal neurons (up to 110 ms), McAllen and Spyer (1978) concluded that it was unlikely that a simple, fast-conducting, disynaptic pathway was involved.

The main efferent sympathetic pathways influencing cardiovascular function arise from bulbospinal neurons which have cell bodies in the medullary reticular vasomotor centres (described above) and whose axons terminate in the intermedio-lateral cell column of the spinal cord. There they synapse (either directly or by means of short interneurons) with the sympathetic pre-ganglionic neurons whose axons run in the thoracolumbar outflow of the autonomic nervous system. There are both inhibitory and excitatory bulbospinal fibres. The inhibitory fibres descend mostly in the lateral (and partly ventral) funiculus of the spinal cord (Coote and Macleod 1974; Henry and Calaresu 1974) and the excitatory fibres descend in the dorsolateral funiculus (Henry and Calaresu 1974).

A great deal of work has been concerned with the nature of the transmitter substances released by the medullary interneurons and descending bulbospinal neurons. Dahlström and Fuxe (1964) described two medullary sites of origin of descending noradrenergic fibres, one corresponding to the vasopressor area in the ventrolateral medulla (A_1), and another corresponding to the NTS and vagal complex (A_2). While the A_2 cells receive direct synaptic input from IXth and Xth cranial nerve fibres (Palkovits and Záborszky 1977), removal of the nodose ganglion caused no terminal degeneration in the A_1 region. However, NTS lesions caused extensive damage at that site (Palkovits and Záborszky 1977), which is further evidence that the baroreceptor information is carried to the 'vasopressor' regions in the medullary reticulum by means of secondary neurons which arise from the NTS. Dahlström and Fuxe (1964) also demonstrated the presence of tryptaminergic terminals, in the NTS and vagal nuclei, which originated mainly from the raphé nucleus. Furthermore, there are neurons containing adrenalin within the medullary areas involved in cardiovascular reflexes (Bolme, Corrodi, Fuxe, Hökfelt, Lidbrink, and Goldstein 1974), and it has been suggested that adrenalin might be a transmitter in the medullary depressor system (Bolme *et al.* 1974).

Although the presence of the various transmitter substances is well established, their relative roles in cardiovascular control are still open to debate. Permanent, electrically-induced, lesions in the NTS of the rat cause severe hypertension, whereas more restricted damage to noradrenergic neurons only causes a transient hypertension (Doba and Reis 1974). Conversely, injection of noradrenalin into the NTS causes hypotension and bradycardia (Struyker-Boudier, Smeets, Brouwer, and van Rossum 1975). Intracisternal administration of 6-hydroxydopamine (which causes a selective degeneration of catecholaminergic nerves, particularly in the spinal cord) prevents the development of the severe hypertension caused by lesions in the NTS (Doba and Reis 1974) but has no long-lasting effect on arterial blood-pressure (Chalmers and Reid 1972). Thus there may be two different catecholaminergic systems involved in the central pathways of the baroreflex: an inhibitory one in the NTS and an excitatory one in the descending bulbospinal pathways (Chalmers 1975).

While there is further evidence in the literature for the existence of an excitatory noradrenergic bulbospinal pathway (Hare, Neumayr, and Franz 1972; Neumayr, Hare, and Franz 1974), certain other findings are not consistent with this. For example, De Groat and Ryall (1967), Ryall (1967), and Coote and Macleod (1974) found that noradrenalin inhibited the activity of sympathetic pre-ganglionic neurons in the lateral columns (Fig. 2.3). There is also disagreement about the role of 5-hydroxytryptamine in the pathways concerned with cardiovascular control. Some workers (Hare *et al.* 1972; Ito and Schanberg 1972; Coote and Macleod 1974; Neumayr *et al.* 1974; Fig. 2.4) have produced evidence of an inhibitory influence of 5-hydroxytryptamine on pre-ganglionic sympathetic nerve activity, whereas others (De Groat and Ryall 1967; Ryall 1967; Wing and Chalmers 1974) claim that it has excitatory effects (Fig. 2.3). At present, no conclusive statements can be made about the medullary centres and their respective transmitters; it is likely that sympathetic vasoconstrictor activity is modulated by a balance between noradrenergic, serotonergic, and adrenergic systems.

If the pathway described above represented the baroreflex system in its entirety then presumably decerebrate, thalamic, and sham-operated animals should all show the same reflex responses to carotid sinus nerve stimulations; but Korner (1975) found that decerebration or hypothalamic ablation decreased the range over which the sinus pressure affected the arterial blood-pressure. Carotid sinus nerve stimulation can alter the firing patterns of populations of neurons in both the anterior hypothalamus (Spyer 1972 and Fig. 2.5) and the posterior

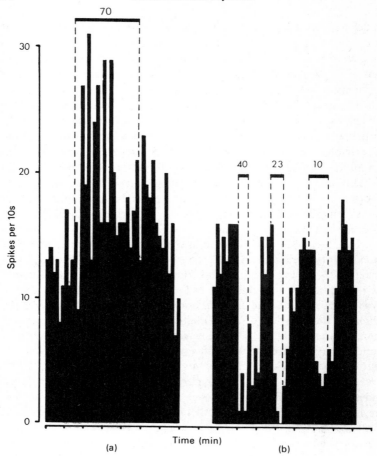

Fig. 2.3. Excitatory action of (*a*) 5-hydroxytryptamine and (*b*) inhibitory effect of noradrenalin on the spontaneous firing of a thoracic pre-ganglionic neuron. The durations of the electrophoretic administrations are indicated by the horizontal bars; numbers represent the strength of the electrophoretic currents in nanoamps (10^{-9} amp). (With permission, Ryall (1967).)

hypothalamus (Thomas and Calaresu 1974) and hypothalamic stimulation can alter some aspects of the baroreflex response (see later).

Like the medulla, the hypothalamus contains pressor (posterior) and depressor (anterior) regions. Stimulation of the anterior regions inhibits sympathetic outflow to the heart and vasculature (Folkow, Langston,

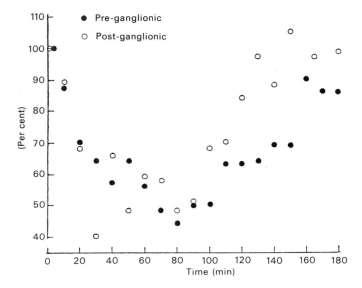

Fig. 2.4. Effect of 5-hydroxytryptophan (a precursor of 5-hydroxytryptamine) on reflex responses in pre-ganglionic and post-ganglionic sympathetic nerves in a decebrate cat spinalized at C1. The reflex responses are expressed as a percentage of the pre-drug value; 5-hydroxytryptophan (i.v. 70 mg/kg) was administered at time = 0. (With permission, Coote and Macleod (1974).)

Öberg, and Prerovsky 1964) whereas stimulation of the posterior regions causes a biphasic rise in blood-pressure, the initial elevation being due to increased sympathetic nerve activity and the secondary rise due to adrenal catecholamine release (Eferakeya and Buñag 1974). Supramedullary areas such as the cingulate gyrus, orbital cortex, and amygdala can also effect cardiovascular function; the pathways from these centres chiefly relay in the hypothalamus.

As mentioned above, hypothalamic stimulation can exert important influences on baroreflex responses; one such effect is seen in the 'defence reaction' which can occur in states of arousal (for full description see Hilton 1966 and Fig. 2.6). The response is due to activity in cortical and limbic structures, and relays through the posterior hypothalamus. The cardiovascular effects include increased cardiac output, increased blood-pressure, and increased heart rate — not the expected reflex bradycardia which should arise from baroreceptor activity (see below). There is vasoconstriction in most of the vascular beds except

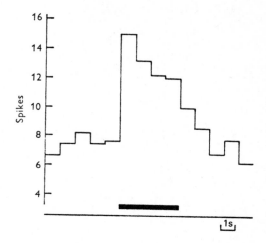

Fig. 2.5. Histogram showing the average response of nine neurons in the anterior hypothalamus exhibiting an increase in discharge frequency on raising the intrasinusal pressure. Pressure within the isolated carotid sinus was raised to 200 mm Hg at the horizontal bar. (With permission, Spyer (1972).)

for skeletal muscle where there is a cholinergically-mediated vasodilatation. Some workers have suggested that hypothalamic activation only inhibits the cardiac vagal part of the baroreflex (Gebber and Snyder 1969; Thomas and Calaresu 1974) whilst others have shown vasoconstrictor tone is also affected (Hilton 1963; Coote and Perez-Gonzalez 1972). The latter findings explain, more fully, the general vasoconstriction which occurs in the defence reaction. The site of inhibition of the baroreflex responses has been investigated. Posterior hypothalamic stimulation inhibits the vagal bradycardia which is normally induced by stimulation of the NTS, although the bradycardia arising from direct stimulation of the nucleus ambiguus is unaffected (Thomas and Calaresu 1974 and Fig. 2.7); this suggests that inhibition of vagal activity occurs in the region of the NTS rather than at the level of the pre-ganglionic efferent neurons. Weiss and Crill (1969) found that posterior hypothalamic stimulation caused depolarization in afferent fibres coming from the carotid sinus, and suggested that the depressant effects on the baroreflex were due to pre-synaptic inhibition at this point. Recently, Jordan and Spyer (1979) were unable to detect any pre-synaptic effects on sinus nerve afferents when conditioning stimuli

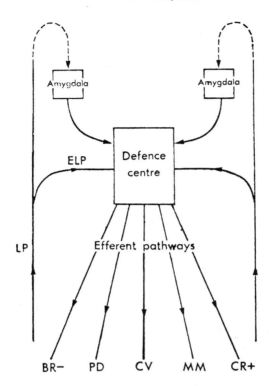

Fig. 2.6. Diagrammatic representation of brainstem centre integrating the 'defence reaction' and its connections. Interrupted line shows possible pathway for conditioned reactions. ELP: extra-lemniscal sensory pathway; LP: lemniscal sensory pathway; BR-: inhibition of baroreceptor reflex; CR+: facilitation of chemoreceptor reflex; CV: cardiovascular response; PD: pupillary dilation; MM: muscular movements. (With permission, Hilton (1966).)

were applied to the hypothalamic defence area, but they did observe effects in glossopharyngeal afferents. Jordan and Spyer (1979) suggested that Weiss and Crill had recorded from glossopharyngeal rather than sinus nerve afferents. The precise mechanism by which hypothalamic stimulation alters baroreflex responses remains to be determined.

Pharmacological investigations have demonstrated that involvement of both α- and β-adrenoceptors in the cardiovascular effects of activity in the higher centres. The pressor response which results from electrical stimulation of the posterior hypothalamus can be blocked by central

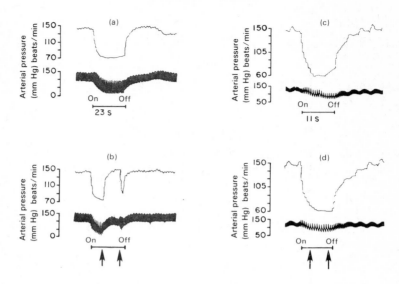

Fig. 2.7. Effect of stimulation of the posteromedial hypothalamus (PMH) on medullary vagal bradycardia elicited by stimulation of the nerve of tractus solitarius (NTS) and n. ambiguus (NA). In each panel: top trace is heart rate in beats/min, bottom trace is arterial pressure in mm Hg; medullary stimulation is represented by solid bar and PMH stimulation occurred between arrows. (a) Stimulation of NTS; (b) simultaneous stimulation of NTS and PMH; (c) stimulation of NA; (d) simultaneous stimulation of NA and PMH. Note that PMH stimulation inhibits NTS bradycardia but not NA bradycardia. (With permission, Thomas and Calaresu (1974).)

α-adrenoceptor antagonism (Przuntek, Guimarães, and Philippu 1971) and enhanced by α-adrenoceptor stimulation (Philippu, Demmeler, and Roensberg 1974). Noradrenalin injected into the cerebral ventricles, in a dose and by a route which principally affects the hypothalamus, can also cause a pressor response; some workers have shown this response to be entirely due to β-adrenoceptor stimulation (e.g. Gagnon and Melville 1966) whilst others have suggested the involvement of both α- and β-adrenoceptors (e.g. Day, Poyser, and Sempik 1976). Borkowski and Finch (1977) found that adrenalin injected into the cerebral ventricles caused bradycardia and a reduction in arterial blood-pressure; they showed that this depressor response was prevented or even reversed by β-adrenoceptor antagonism but was unaffected by α-adrenoceptor blockade. However, the depressor response induced by injection of noradrenalin directly into the anterior hypothalamus (Struyker-Boudier et al. 1975) was prevented by α-adrenoceptor blockade and was unaffected by the administration of β-adrenoceptor antagonists. This evidence collectively suggests the involvement of α- and β-adrenoceptors in both the pressor and depressor responses to hypothalamic stimulation.

Hypothalamic stimulation can also affect cardiovascular control mechanisms through hormonal changes, independent of the influence exerted on baroreflexes. Releasing factors secreted by the hypothalamus can lead to the release of a number of hormones from the anterior pituitary (for example, adrenocorticotrophic hormone (ACTH); follicle-stimulating hormone; luteinizing hormone; thyrotrophin; and growth hormone). Of these, ACTH in particular can have important, albeit indirect, effects on the cardiovascular system due to its stimulatory action on the adrenal cortex.

Besides their peripheral cardiovascular effects there is evidence that pituitary peptides such as ACTH and vasopressin may influence the integrative control of the cardiovascular system; for example, Bohus (1974) showed that the cardiovascular response to posterior hypothalamic stimulation was markedly altered in hypophysectomized rats treated with either vasopressin or ACTH. Presently there is also a great deal of interest in the possible involvement of neuropeptides as modulators of activity in neuronal systems concerned with cardiovascular control. Compounds such as Substance P, neurotensin, met-enkephalin, and various others may turn out to be of central importance in cardiovascular homeostasis.

2.1.3. Baroreceptor reflexes

Heymans and Neil (1958) reviewed data from numerous studies involving carotid artery occlusion, sinus and aortic nerve section or stimulation and concluded that there was a tonic discharge from the arterial baroreceptors which inhibited the spontaneously active medullary cardiovascular centres. After occlusion of the carotid artery proximal to the sinus, the fall in intrasinus pressure reduces the distension of the walls and hence inhibits baroreceptor afferent discharge. This permits the spontaneous activity of the medullary centres to be expressed and there is an increased efferent discharge in cardiac sympathetic and sympathetic vasomotor nerve fibres (to different extents in different vascular beds — see below) which tends to increase blood-pressure. If the aortic nerves are sectioned prior to carotid artery occlusion, the rise in blood-pressure is greater — presumably due to removal of an inhibitory influence from aortic arch baroreceptor afferents which would be stimulated by the rise in systemic arterial pressure.

When the carotid sinus and aortic arch are simultaneously exposed to increased transmural pressure there is increased afferent discharge which may inhibit sympathetic vasomotor and excite cardiac vagal outflow. There is some, albeit equivocal, evidence that baroreceptor afferent discharge can also elicit vasodilatation by activating cholinergic vasodilator nerves (Uvnäs 1966; Takeuchi and Manning 1971).

Bradycardia occurs in response to a rise in systemic arterial blood-pressure in anaesthetized and conscious states, but the efferent activity responsible for the fall in heart rate appears to differ in the two states. Vatner, Franklin, and Braunwald (1971) found that the bradycardia was prevented by propranolol (i.e. mediated by sympathetic withdrawal) in anaesthetized dogs, but was blocked by atropine (i.e. due to vagal discharge) in conscious animals. As Kirchheim (1976) has pointed out, the difference between the responses in anaesthetized and conscious animals probably depends upon general anaesthesia interfering with baroreceptor afferent signalling to vagal cardiac efferents and also upon changes in various afferent inputs due to surgical procedures in the anaesthetized animals not employed at the time of recording from the conscious animals.

Under conditions where anaesthesia and surgery are not complicating factors, the afferent inputs from carotid sinus and aortic arch have differential effects on heart rate. For example, in conscious man electrical stimulation of the carotid sinus nerves has only moderate effects on heart rate (Epstein, Beiser, Goldstein, Stampfer, Wechsler, Glick, and Braunwald 1969), whereas manoeuvres which also stimulate aortic arch receptors have much more marked effects (Mancia, Ferrari, Gregorini, Valentini, Ludbrook, and Zanchetti 1977 and Fig. 2.8).

There has been much discussion about the reflex effects of stimulation of carotid sinus or aortic arch baroreceptors on vascular resistance (Angell-James and Daly 1970; Dampney, Taylor, and McLachlan 1971; Donald and Edis 1971). Donald and Edis (1971) monitored perfusion pressure in a hind-limb and related the change in this variable to carotid sinus or aortic arch pressure. They found that the stimulus–response

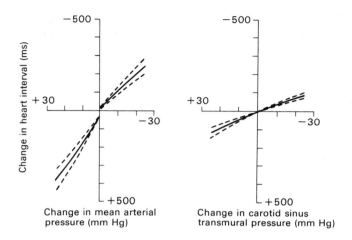

Fig. 2.8. Changes in heart interval with drug-induced changes in mean arterial pressure (left) and with neck chamber-induced changes in carotid sinus transmural pressure (right); means (continuous lines) ± SE (dashed lines) of individual regression coefficients taken from the eight subjects in which both techniques were used. For the neck chamber the early heart interval responses were considered, the change in carotid sinus transmural pressure being that at the time at which the response was developed. (With permission, Mancia et al. (1977).).

curve for the aortic arch had a higher threshold and a lower sensitivity than that for the carotid sinus (Fig. 2.9). Dampney et al. (1971) suggested such differences might be due to differences in the baroreceptor afferents themselves. Such an assertion is consistent with the findings of Pelletier, Clement, and Shepherd (1972) who compared afferent activity in multifibre preparations of canine sinus and aortic nerves. Measurements of mean impulse activity versus mean systemic arterial pressure indicated that aortic baroreceptors had a higher threshold and lower sensitivity than those of the carotid sinus (Pelletier et al. 1972). Recently, Samodelov, Godehard, and Arndt (1979) examined this possibility in decerebrated cats. In 11 cases they were able to record simultaneously from single-fibre preparations of carotid sinus and aortic arch baroreceptors in the same animal. Samodelov et al. (1979) could find no evidence of differences in sensitivity or working range between carotid sinus and aortic arch baroreceptors; thus any differential actions of carotid sinus and aortic arch afferents on vascular

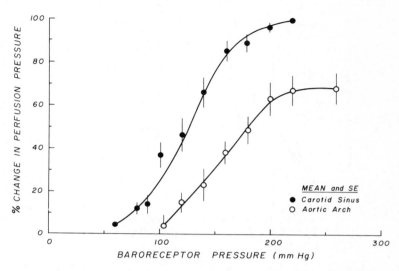

Fig. 2.9. Stimulus response curves for the carotid sinus and aortic arch baroreflexes in eleven dogs. Response in hind-limb perfusion pressure is expressed as percentage of maximal change evoked by carotid distension (mean ± SE). The aortic arch curve was displaced to the right, and its maximal slope and height were less than those for the carotid sinus curve. (With permission, Donald and Edis (1971).)

resistance are likely to be due to differences in central connections or effects.

However, there is not unanimous agreement that carotid sinus and aortic arch baroreceptors play different roles in the control of systemic arterial blood-pressure. Ito and Scher (1978) found, in dogs, that section of the carotid sinus nerves did not affect mean arterial blood-pressure when the animals were monitored over long periods of time in the conscious state (Fig. 2.10). They therefore concluded that the aortic arch baroreceptors were competent to adjust for acute increases and decreases in systemic arterial blood-pressure. These workers subsequently (Ito and Scher 1979) showed that denervation of aortic baroreceptors caused hypertension in unanaesthetized dogs with carotid sinus nerves intact. There are no ready explanations for the inconsistent results in this area (compare Cowley, Quillen, and Barber 1980).

While many experiments have been carried out on the cardiovascular effects of carotid sinus and aortic nerve stimulation, this does not permit an analysis of the relative contributions of different afferent

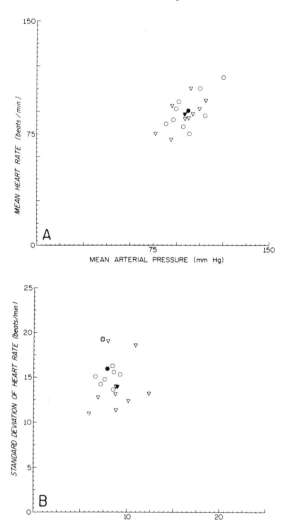

Fig. 2.10. A: Comparison of mean values of arterial pressure and heart rate in nine dogs before (○) and after (∇) section of the carotid sinus nerves. The solid symbols indicate the mean values for all dogs before and after nerve section. B: Standard deviations of beat mean arterial pressure and heart rate for all recording sessions before and after section of the carotid sinus nerves from the same nine dogs. (With permission, Ito and Scher (1978).)

fibre-types to the reflex responses seen. Thorén and Jones (1977) reported that aortic baroreceptor C-fibres in the rabbit had a higher threshold and were less sensitive than the B-fibres and suggested that the former might play an important homeostatic role at high pressure (Fig. 2.11). Aars, Myhre, and Haswell (1978) found that a rise of 20–30 mm Hg in mean systemic arterial pressure caused a 60 per cent inhibition of renal nerve activity (in chloralose-urethane anaesthetized rabbits)

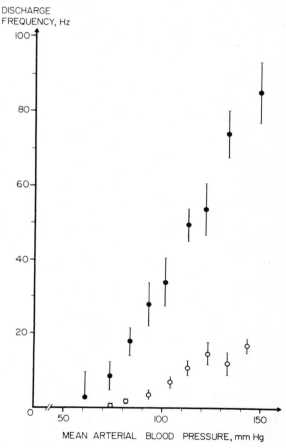

Fig. 2.11. Mean (± SEM) stimulus response curves for 18 C-fibres (○) and 10 medullated fibres (●) in the aortic nerve of the rabbit. (With permission, Thorén and Jones (1977).).

when the aortic nerves were intact. Similar pressure changes when the B-fibres had been inactivated by anodal block elicited no reflex effects. This indicates that the C-fibres were not activated, since direct electrical stimulation of aortic nerve C-fibres caused profound inhibition of renal nerve activity. Like Thorén and Jones (1977), these workers concluded that the C-fibres acted as a brake mechanism at high pressures.

It is not clear to what extent activation of different populations of afferent fibres might be responsible for the differential reflex effects of arterial baroreceptor stimulation on individual vascular beds. Kendrick, Öberg, and Wennergren (1972) found, in the cat, that changes in arterial baroreceptor activity had a twofold greater influence on vasoconstrictor discharge to skeletal muscle than to the kidney; vasoconstrictor discharge to the gut was influenced to about the same extent as that to muscle, while there was no observable effect on vasoconstrictor activity affecting cutaneous arteriovenous anastomoses, although the cutaneous circulation proper was affected to the same extent as that in muscle (Fig. 2.12). In man, carotid stimulation has marked influences on muscle and skin resistance vessels (Beiser, Zelis, Epstein, Mason, and Braunwald 1970).

There is some disagreement about the involvement of veins in

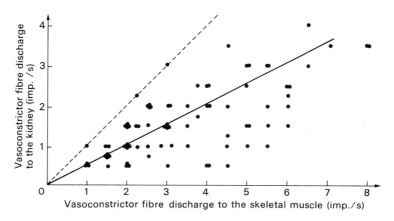

Fig. 2.12. Relationship between the average discharge rates in calf muscle and renal vasoconstrictor fibres attained with reductions of baroreceptor activity. The dots indicate individual results from 81 tests in 14 animals. The heavy line is the regression line, passing through the origin. ($y = 0.4972\ x$). Broken line is the 'identity' line. (With permission, Kendrick *et al.* (1972).)

baroreceptor reflexes. Epstein *et al.* (1969) concluded that carotid sinus nerve stimulation in man had no effect on any venous bed, but Shepherd and Vanhoutte (1978) indicate that the splanchnic venous system is likely to be involved in arterial baroreflexes (Fig. 2.13). In the dog, changes in carotid sinus pressure have important effects on splanchnic venous capacitance (Hainsworth and Karim 1976 and Fig. 2.14). Apart from the effects of the cardiovascular system mentioned above, arterial baroreceptors also affect renal function in a way which can influence long-term cardiovascular homeostasis (see pp. 98; 119).

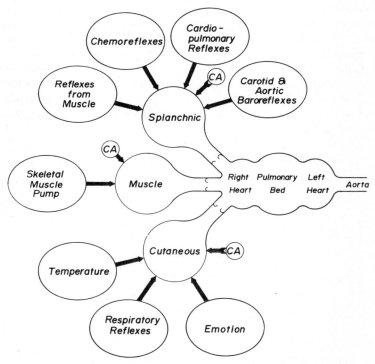

Fig. 2.13. Regulation of central blood volume and cardiac filling pressure by major components of systemic venous system — splanchnic, muscle, and cutaneous veins. Active changes in venous capacity are caused by contraction or relaxation of smooth muscle in venous walls. Whereas splanchnic and cutaneous veins have an abundance of smooth muscle and a rich sympathetic innervation, muscle veins have little smooth muscle and few if any sympathetic nerves. Skeletal muscle pump has a major role in reducing capacity of muscle veins. CA = circulatory catecholamines. (With permission, Shepherd and Vanhoutte (1978).)

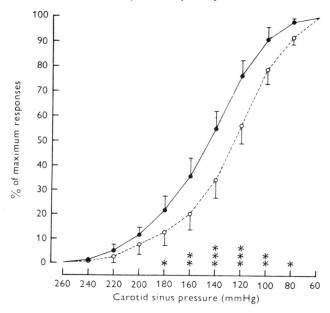

Fig. 2.14. Responses of abdominal vascular resistance (○) and capacitance (●) at different carotid pressures. Results of means ± 1 SE of mean from ten dogs. The significance of differences between resistance and capacitance responses at the various carotid pressures is given by the asterisks: *$P<0.05$; **$P<0.01$; ***$P<0.005$ (paired t-tests). (With permission, Hainsworth and Karim (1976).)

2.2. CARDIOPULMONARY RECEPTORS

2.2.1. Atrial receptors with vagal afferents

Various types of atrial receptors with vagal afferents have been distinguished on the basis of the patterns of action-potential discharge in their associated axons (Paintal 1973). Paintal (1973) described type-A receptors, whose discharge pattern was largely associated with the 'a' wave of the atrial pressure curve, type-B receptors which discharged action potentials during the 'v' wave, and intermediate receptors which had some characteristics of both A- and B-receptors. All these receptors are associated with myelinated nerve fibres, but atrial receptors associated with unmyelinated C-fibres have also been described (Coleridge, Coleridge, Dangel, Kidd, Luck, and Sleight 1973; Thorén 1976a). Some of these atrial C-fibre afferents show action potential discharge patterns similar to those described by Paintal (1973).

There is no histological evidence for the existence of distinct receptor-types, and the possibility exists that the different firing patterns observed are due to different locations of the afferent endings in the atrial walls. Strong support for this belief comes from the finding that the discharge patterns of receptors can be converted by manoeuvres which affect atrial behaviour during the cardiac cycle (Kappagoda, Linden, and Mary 1976, 1977a, b and Fig. 2.15). However, although there may not be distinct receptor-types, it does not follow that the different sorts of afferent information have the same significance centrally.

The main problem associated with elucidating the reflex effects of stimulating atrial receptors is due to the difficulty of localizing the stimulus. Intravenous infusion of fluids, haemorrhage, or distension of balloons in the atria will stimulate atrial receptors, but may also cause complex haemodynamic changes affecting receptors in various other parts of the cardiovascular system. Linden and his colleagues (see Linden 1975) have devised a technique whereby stretching of a circumscribed region is achieved by inflating small balloons (usually located at the atriovenous junctions). While this manoeuvre certainly stimulates atrial endings with myelinated afferents, the possible contribution of endings with unmyelinated afferents to the reflex responses seen has not been systematically investigated (see Thorén 1979). Linden and his colleagues presently believe that stimulation of endings in left or right atria causes a moderate increase in heart rate due to afferent activity in the vagi and efferent activity in the cardiac sympathetic nerves. A possible contribution of vagal inhibition to the tachycardia has been claimed (Burkhart and Ledsome 1974) but not confirmed (Kappagoda, Linden, and Mary 1975). It has been reported that atrial distension can cause tachycardia or bradycardia, depending on the initial heart rate (Edis, Donald and Shepherd 1970), but in these experiments the technique used could also have caused stimulation of other receptors (Linden 1975). The efferent fibres activated by atrial stimulation appear to innervate the sino-atrial node only, since the increases in heart rate were unaccompanied by inotropic effects (Linden 1975).

There is some evidence that circumscribed distension of pulmonary vein–atrial junctions causes a reflex vasodilatation localized to the renal vasculature (Mason and Ledsome 1974 and Fig. 2.16) which is consistent with the inhibition of renal nerve activity seen under these

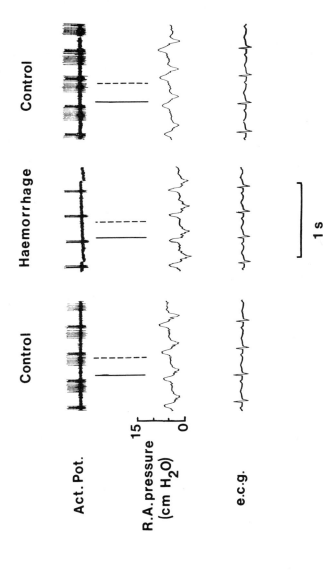

Fig. 2.15. Conversion of the pattern of discharge in an intermediate-type receptor. From above downwards record of action potentials, right atrial pressure and electrocardiogram (e.c.g.). The vertical lines are drawn to show the temporal relationship between bursts of action potentials and waves of the atrial pressure pulse. The continuous line indicates the end of the 'a' wave and the interrupted line indicates the peak of the 'v' wave of the atrial pressure pulse. *Left*: initial control record; *centre*: after bleeding 140 ml; *right*: final control record. In the control periods the unit behaved as an intermediate-type receptor and during the haemorrhage it behaved as a type-A receptor. (With permission, Kappagoda *et al.* (1977).)

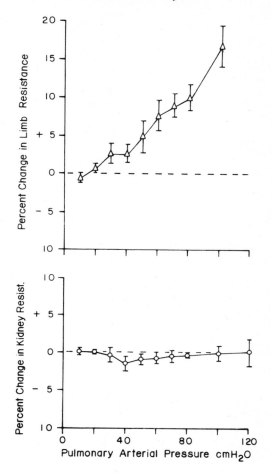

Fig. 2.16. Changes in hind-limb resistance and renal resistance in response to changes in perfusion pressure in a pulmonary arterial pouch. Results for limb resistance are average changes from values at a pulmonary arterial pouch pressure of 18 cm H_2O in six dogs ± SEM. Results for renal resistance are calculated similarly in six dogs. (With permission, Mason and Ledsome (1974).)

conditions (Karim, Kidd, Malpus, and Penna 1972 and Fig. 2.17). Manoeuvres which cause more widespread afferent stimulation elicit skeletal muscle vasomotor effects (Mason and Ledsome 1974), although some findings indicate that such effects might arise from atrial receptor

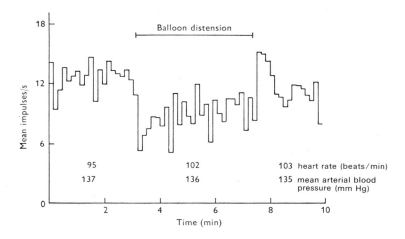

Fig. 2.17. Continuous record of the mean frequency of action potential discharge in a renal sympathetic efferent strand. Balloon distension of the pulmonary vein − left atrial junction resulted in a significant decrease in the mean frequency without significant changes in mean arterial blood pressure, or heart rate. (With permission, Karim *et al.* (1972).)

influences (Mancia and Donald 1975). One possible explanation for the various effects observed is that under the different experimental conditions, varying proportions of myelinated and unmyelinated afferents were being stimulated (Thorén 1979). Whatever the reason for these differences, as Öberg (1976) has pointed out, the over-all circulatory effects elicited from atrial receptors are modest and it may be that these receptors are more importantly concerned in modulating renal function (see pp. 98; 119). Such an assertion is consistent with disordered renal function in conditions where afferent signalling from atrial receptors is likely to be abnormal (Zucker, Earle, and Gilmore 1979).

2.2.2. Ventricular receptors with vagal afferents

In addition to atrial receptors, the heart contains mechanoreceptors situated in the ventricles. There are two types of ventricular receptor; one with a pulsatile discharge in time with the heartbeat and the other with an irregular discharge pattern. The former type are few in number; they are thought to be stimulated by ventricular contraction and/or a change in pressure in the ventricles. The reflex cardiovascular responses evoked by moderate rises in ventricular pressure are, in some ways, similar to those which follow activation of arterial baroreceptors. There is a vagally-mediated bradycardia and vasodilatation in most vascular beds, due to inhibition of sympathetic vasoconstrictor activity; these effects are attributable to stimulation of the ventricular receptors with myelinated afferents. The receptors which discharge in an irregular manner have non-myelinated afferent fibres; they are probably stimulated during vigorous ventricular contraction (Sleight and Widdicombe 1965). Moderate changes in pressure cannot be sensed by these receptors and it is possible that, rather than being involved in the short-term regulation of blood-pressure, they are primarily engaged in the regulation of blood volume. Ventricular receptors associated with unmyelinated afferents can be stimulated with drugs such as nicotine or veratridine; the responses to such stimuli are a reflex hypotension and bradycardia — the classic Bezold–Jarisch effect. These receptors also appear to be stimulated by myocardial ischaemia caused by coronary artery occlusion (Thorén 1976b). Thames and Abboud (1979) have shown that the reflex responses to circumflex coronary artery occlusion in the dog are due to stimulation of left ventricular receptors with vagal afferents. The input from these receptors appears to override that from arterial baroreceptors, since when they are stimulated by coronary artery occlusion there is a reflex inhibition of renal sympathetic nerve activity, even though systemic arterial blood-pressure falls. While the function of these ventricular receptors is obscure at present, various workers have suggested they might be responsible for some of the cardiovascular sequelae of acute myocardial infarction in man. The finding that patients with acute myocardial infarction and elevated left-sided filling pressures have disordered renal function (Bennett, Brooke, Lis, and Wilson 1979) is consistent with the experimental findings.

2.2.3. Cardiopulmonary receptors with sympathetic afferents

There is evidence for the existence of atrial and ventricular receptors whose afferent fibres run in the cardiac sympathetic nerves (Brown 1979). The majority (about 70 per cent) of these afferent fibres are myelinated and fall into the Aδ category; the remainder are C-fibres.

Like the cardiac receptors with vagal afferents, the myelinated sympathetic afferents show cardiac rhythmicity in their discharges, while the unmyelinated afferents show irregular discharges under normal conditions. The general finding is that electrical stimulation of cardiac sympathetic afferents has positive chronotropic and inotropic effects (due to inhibition of efferent vagal activity and enhancement of efferent sympathetic effects), associated with peripheral vasoconstriction (particularly in the kidney). Since stimulation of cardiac sympathetic afferents can inhibit vagal efferent effects it is clear that these reflexes have medullary components. While there is some evidence that cardiopulmonary sympathetic afferents may be stimulated by coronary artery occlusion, their role in cardiovascular homeostasis awaits investigation.

2.2.4. Pulmonary receptors

There are three main types of receptor situated in either the lungs or the thoracic wall. Type-J (juxtacapillary) receptors are associated with the pulmonary vasculature and may be stimulated by an increased interstitial pressure or volume; the reflex response is tachycardia and dyspnoea (Paintal 1969). Stretch receptors located mainly in the respiratory bronchioles show a sustained discharge, the frequency of which increases in proportion to the depth of inspiration; these induce tachycardia and vasodilatation when stimulated by lung inflation. The other type of pulmonary baroreceptor is associated with the trachea and extrapulmonary bronchi; stimulation of this type of receptor as a result of deep breathing or hyperventilation evokes a reflex arteriolar and venous constriction (Eckstein, Hamilton, and McCammond 1958).

The majority of the systems mentioned above serve to maintain systemic arterial pressure within certain limits. However, if the blood-pressure falls below a critical level, arterial baroreceptors cease to signal and so can have no greater influence on efferent output. If the fall in systemic arterial pressure is such that tissues are inadequately perfused then the reductions in arterial P_{O_2} and pH and the increase in P_{CO_2} will cause chemoreceptor stimulation. Any reduction in chemoreceptor perfusion due to reflex constriction of the vessels supplying these structures would augment these effects (Öberg 1976).

2.3. CHEMORECEPTORS

Peripheral chemoreceptors are largely located in the carotid and aortic bodies and, while stimulation of them can have profound cardiovascular effects, their role in cardiovascular homeostasis under normal conditions has been contended. As Daly (1972) has pointed out, the confusion has probably arisen from the fact that chemoreceptor stimulation can elicit a variety of responses which may interact in such a way as to obscure the direct reflex effects.

When experiments are carried out in such a way that the primary responses to carotid chemoreceptor stimulation can be examined, there is a bradycardia (largely mediated by increased vagal efferent activity), and a sympathetically-mediated constriction in skin, muscle, splanchnic, and pulmonary vascular beds (see Daly 1972). Heistad, Abboud, Mark, and Schmid (1975), however, observed vasodilatation in the dog's paw in response to chemoreceptor stimulation. They argued that this represented a cutaneous vasodilatation, and produced evidence that it was mediated by a non-adrenergic, non-cholinergic mechanism. However, Sybertz and Zimmerman (1977) found that the response was inhibited by guanethidine, indicating that the dilatation was due to inhibition of noradrenergic vasoconstrictor tone.

Heistad *et al.* (1975) suggested that redirection of blood from muscle to skin might have an important role in oxygen conservation during the chemoreceptor reflex. The usefulness of cutaneous hyperperfusion is not apparent, although increased coronary blood-flow would clearly be appropriate; Abboud, Heistad, Mark, and Schmid (1975) have shown that chemoreceptor stimulation causes a reflex activation of vagal, cholinergic, vasodilator fibres to the coronary vessels. In the conscious state the vasodilatation is augmented by noradrenergic vasoconstrictor withdrawal (Vatner and Braunwald (1975).

Daly (1972) has suggested that the primary reflex response to chemoreceptor stimulation may play an important role in maintaining blood-pressure after a haemorrhage, and these responses may be enhanced by arterial baroreceptor unloading (Heistad, Abboud, Mark, and Schmid 1974). However, carotid chemoreceptor stimulation in spontaneously respiring animals causes hyperventilation, and hyperventilation produces tachycardia and peripheral vasodilatation (due to a fall in arterial P_{CO_2} and activation of the lung inflation reflex (see above; Daly 1972)). The observation that conscious animals when haemorrhaged respond with peripheral vasoconstriction (Vatner and Braunwald 1975) would seem to indicate that under these conditions the chemoreceptors are not stimulated to any great extent.

There is much discussion about the mechanisms concerned in the acute and chronic control of systemic arterial blood-pressure. Presently it would be fair to say that there is a rift between those who believe that the long-term control of blood-pressue is dependent on renal function (Guyton, Coleman, Cowley, Manning, Norman, and Ferguson 1974) and those who believe that various cardiovascular reflexes are likely to influence blood-pressure both acutely and in the long term (see Zanchetti 1979). However, the importance of renal function in cardiovascular homeostasis cannot be denied, and the following sections consider those aspects of salt and water balance that are thought to contribute to the control of the volume and osmolality of extracellular fluid.

REFERENCES

Aars, H., Myhre, L., and Haswell, B. A. (1978). The function of baroreceptor C fibres in the rabbit's aortic nerve. *Acta physiol. Scand.* **102**, 84–93.

Abboud, F. M. (1979). Integration of reflex responses in the control of blood pressure and vascular resistance. *Amer. J. Cardiol.* **44**, 903–11.

—— Heistad, D. D., Mark, A. L., and Schmid, P. G. (1975). Differential responses of the coronary circulation and other vascular beds to chemoreceptor stimulation. In *The peripheral arterial chemoreceptors* (ed. M. J. Purves), pp. 427–42. Cambridge University Press.

Angell-James, J. E. (1971). The effects of changes of extramural, "intrathoracic" pressure on aortic arch baroreceptors. *J. Physiol.* **214**, 89–103.

—— and Daly, M. de B. (1970). Comparison of the reflex vasomotor responses to separate and combined stimulation of the carotid sinus and aortic arch baroreceptors by pulsatile and non-pulsatile pressures in the dog. *J. Physiol.* **209**, 257–93.

Beiser, G. D., Zelis, R., Epstein, S. E., Mason, D. T., and Braunwald, E. (1970). The role of skin and muscle resistance vessels in reflexes mediated by the baroreceptor system. *J. clin. Invest.* **49**, 225–31.

Bennett, E. D., Brooke, N., Lis, Y., and Wilson, A. (1979). Is the elevated renal function in patients with acute heart failure a homeostatic mechanism? In *Cardiac receptors* (ed. R. Hainsworth, C. Kidd, and R. J. Linden), p. 468. Cambridge University Press.

Bohus, B. (1974). The influence of pituitary peptides on brain centers controlling autonomic responses. *Prog. Brain Res.* **41**, 175–83.

Bolme, P., Corrodi, H., Fuxe, K., Hökfelt, T., Lidbrink, P., and Goldstein, M. (1974). Possible involvement of central adrenaline neurons in vasomotor and respiratory control. Studies with clonidine and its interactions with piperoxane and yohimbine. *Eur. J. Pharmacol.* **28**, 89-94.

Borkowski, K. R. and Finch, L. (1977). Cardiovascular responses to centrally administered adrenaline in spontaneous hypertensive rats. *Brit. J. Pharmacol.* **61**, 130 P.

Brown, A. M. (1979). Cardiac reflexes. In *Handbook of physiology* (ed. R. M. Berne, N. Sperelakis, and S. N. Geiger), Section 2, Vol. 1, pp. 677-90. American Physiology Society, Bethesda, Maryland.

Burkhart, S. M. and Ledsome, J. R. (1974). The response to distension of the pulmonary vein-left atrial junctions in dogs with spinal section. *J. Physiol.* **237**, 685-700.

Chalmers, J. P. (1975). Neuropharmacology of central mechanisms regulating pressure. In *Central action of drugs in blood pressure regulation* (ed. D. S. Davies and J. L. Reid), pp. 36-61. Pitman Medical, London.

—— and Reid, J. L. (1972). Participation of central noradrenergic neurons in arterial baroreceptor reflexes in the rabbit. *Circulation Res.* **31**, 789-804.

Chiba, T. and Doba, N. (1975). The synaptic structure of catecholaminergic axon varicosities in the dorso-medial portion of the nucleus tractus solitarius of the cat: possible roles in the regulation of cardiovascular reflexes. *Brian Res.* **84**, 31-46.

Coleridge, H. M., Coleridge, H. C. G., Dangel, A., Kidd, C., Luck, J. C., and Sleight, P. (1973). Impulses in slowly conducting vagal fibers from afferent endings in the veins, atria and arteries of dogs and cats. *Circulation Res.* **33**, 87-97.

Coote, J. H. and MacLeod, V. H. (1974). The influence of bulbospinal monoaminergic pathways on sympathic nerve activity. *J. Physiol.* **241**, 453-75.

—— and Perez-Gonzalez, J. F. (1972). The baroreceptor reflex during stimulation of the hypothalamic defence region. *J. Physiol.* **224**, 74P-75P.

Cottle, M. K. (1964). Degeneration studies of primary afferents of IXth and Xth cranial nerves in the cat. *J. comp. Neurol.* **122**, 329-43.

—— and Calaresu, F. R. (1975). Projections from the nucleus and tractus solitarius in the cat. *J. comp. Neurol.* **161**, 143-58.

Cowley, A. W. Jr., Quillen, E. W., and Barber, B. J. (1980). Further evidence for lack of baroreceptor control of long-term level of arterial pressure. In *Arterial baroreceptors and hypertension* (ed. P. Sleight), pp. 391-9. Oxford University Press.

Dahlström, A. and Fuxe, K. (1964). Evidence for the existence of monoamine neurones in the central nervous system. I. Demonstration of monoamines in the cell bodies of brain stem neurons. *Acta physiol. Scand.* **62** (Suppl. 232), 1-55.

References

Daly, M. de B. (1972). Interaction of cardiovascular reflexes. In *The scientific basis of medicine. Annual reviews* (ed. I. Gilliland and J. Francis), pp. 307-32. Athlone Press, University of London.

Dampney, R. H. L., Taylor, M. G., and McLachlan, E. M. (1971). Reflex effects of stimulation of carotid sinus and aortic baroreceptors on hindlimb vascular resistance in dogs. *Circulation Res.* 29, 119-27.

Day, M. D., Poyser, R. H., and Sempik, J. (1976). Pressor responses to noradrenaline administered into the third cerebral ventricle of anaesthetized and conscious cats. *Brit. J. Pharmacol.* 57, 450P.

De Groat, W. C. and Ryall, R. W. (1967). An excitatory action of 5-hydroxytryptamine on sympathetic preganglionic neurones. *Exp. Brain Res.* 3, 299-305.

Doba, N. and Reis, D. J. (1974). Role of central and peripheral adrenergic mechanisms in neurogenic hypertension produced by brain stem lesions in rat. *Circulation Res.* 34, 293-301.

Donald, D. E. and Edis, A. J. (1971). Comparison of aortic and carotid baroreflexes in the dog. *J. Physiol.* 215, 521-38.

Eckstein, J. W., Hamilton, W. K. and McCammond, J. M. (1958). Pressure-volume changes in forearm veins of man during hyperventilation. *J. clin. Invest.* 37, 956-61.

Edis, A. J., Donald, D. E., and Shepherd, J. T. (1970). Cardiovascular reflexes from stretch of pulmonary vein-atrial junctions in the dog. *Circulation Res.* 27, 1091-100.

Eferakeya, A. and Buñag, R. D. (1974). Adrenomedullary pressor responses during posterior hypothalamic stimulation. *Amer. J. Physiol.* 227, 114-18.

Epstein, S. E., Beiser, G. D., Goldstein, R. E., Stampfer, M., Wechsler, A. C., Glick, G., and Braunwald, E. (1969). Circulatory effects of electrical stimulation of the carotid sinus nerves in man. *Circulation* 40, 269-76.

Folkow, B. and Neil, E. (1971). *Circulation.* Oxford University Press, New York.

—— Langston, J., Öberg, B., and Prerovsky, I. (1964). Reactions of the different series-coupled vascular sections upon stimulation of the hypothalamic sympatho-inhibitory area. *Acta physiol. Scand.* 61, 476-83.

Gagnon, D. J. and Melville, K. I. (1966). Further observations on the possible role of noradrenaline in centrally mediated cardiovascular responses. *Revue Can. Biol.* 25, 99-105.

Garcia, M., Jordan, D., and Spyer, K. M. (1979). Identification of aortic nerve cell bodies in the nodose ganglion of the rabbit and their central projections. *J. Physiol.* 290, 23P-24P.

Gebber, G. L. and Snyder, D. W. (1969). Hypothalamic control of baroreceptor reflexes. *Amer. J. Physiol.* 218, 124-31.

Gunn, C. G., Sevelius, G., Puiggari, M. J., and Myers, F. K. (1968). Vagal cardiomotor mechanisms in the midbrain of the dog and cat. *Amer. J. Physiol.* 214, 258-62.

Guyton, A. C., Coleman, T. G., Cowley, A. W., Manning, R. D., Norman, R. A., and Ferguson, J. D. (1974). A systems analysis approach to understanding long-range arterial blood pressure control and hypertension. *Circulation Res.* **35**, 159–76.

Hainsworth, R. and Karim, F. (1976). Responses of abdominal vascular capacitance in the anaesthetized dog to changes in carotid sinus pressure. *J. Physiol.* **262**, 659–77.

Hare, B. D., Neumayr, R. J., and Franz, D. N. (1972). Opposite effects of L-Dopa and 5-HTP on spinal sympathetic reflexes. *Nature, (Lond.)* **239**, 336–7.

Hauss, W. H., Kreuziger, H., and Asteroth, H. (1949). Über die Reizung der Pressorezeptoren in Sinus caroticus biem Hund. *Z. Kreislaufforsch.* **38**, 28–33.

Heistad, D. D., Abboud, F. M., Mark, A. L., and Schmid, P. G. (1974). Interaction of baroreceptor and chemoreceptor reflexes – modulation of the chemoreceptor reflex by changes in baroreceptor activity. *J. clin. Invest.* **53**, 1226–36.

——, ——, ——, —— (1975). Response of muscular and cutaneous vessels to physiologic stimulation of chemoreceptors. *Proc. Soc. exp. Biol. Med.* **148**, 198–202.

Henry, J. L. and Calaresu, F. R. (1974). Pathways from medullary nuclei to spinal cardioacceleratory neurons in the cat. *Exp. Brain Res.* **20**, 505–14.

Heymans, C. and Neil, E. (1958). *Reflexogenic areas of cardiovascular system.* Churchill, London.

Hilton, S. M. (1963). Inhibition of baroreceptor reflexes on hypothalamic stimulation. *J. Physiol.* **165**, 56P–57P.

—— (1966). Hypothalamic regulation of the cardiovascular system. *Brit. med. Bull.* **22**, 243–48.

Ito, A. and Schanberg, S. M. (1972). Central nervous system mechanisms responsible for blood pressure elevation induced by *p*-chlorophenylalanine. *J. Pharmacol. exp. Ther.* **181**, 65–74.

Ito, C. S. and Scher, A. M. (1978). Regulation of arterial blood pressure by aortic baroreceptors in the unanaesthetized dog. *Circulation Res.* **42**, 230–6.

——, —— (1979). Hypertension following denervation of aortic baroreceptors in unanaesthetized dogs. *Circulation Res.* **45**, 26–34.

Jordan, D. and Spyer, K. M. (1977). Studies on the termination of sinus nerve afferents. *Pflügers' Arch.* **369**, 65–73.

——, —— (1979). Studies on the excitability of sinus nerve afferent terminals. *J. Physiol.* **297**, 123–34.

Kappagoda, C. T., Linden, R. J., and Mary, D. A. S. G. (1975). Reflex increases in heart rate from stimulation of left atrial receptors. *J. Physiol.* **244**, 78P–79P.

——, ——, —— (1976). Atrial receptors in the cat. *J. Physiol.* **262**, 431–46.

Kappagoda, C. T., Linden, R. J., and Mary, D. A. S. G. (1977a). Patterns of discharge from atrial receptors in the dog. *J. Physiol.* **270**, 65P-66P.

——, ——, —— (1977b). Atrial receptors in the dog and the rabbit. *J. Physiol.* **272**, 799-815.

Karim, F., Kidd, C., Malpus, C. M., and Penna, P. E. (1972). The effects of stimulation of the left atrial receptors on sympathetic efferent nerve fibres. *J. Physiol.* **227**, 243-60.

Keith, I. C., Kidd, C., Malpus, C. M., and Penna, P. E. (1974). Reduction of baroreceptor impulse activity by sympathetic nerve stimulation. *J. Physiol.* **238**, 61P-62P.

Kendrick, E., Öberg, B., and Wennergren, G. (1972). Vasoconstrictor fibre discharge to skeletal muscle, kidney, intestine and skin at varying levels of arterial baroreceptor activity in the cat. *Acta physiol. Scand.* **85**, 464-76.

Kirchheim, H. R. (1976). Systemic arterial baroreceptor reflexes. *Physiol. Rev.* **56**, 100-76.

Korner, P. I. (1971). Integrative neural cardiovascular control. *Physiol. Rev.* **51**, 312-67.

—— (1975). Central and peripheral 'resetting' of the baroreceptor system. *Clin. exp. Pharmacol. Physiol.* Suppl. 2, 171-8.

Kunze, D. L. and Brown, A. M. (1978). Sodium sensitivity of baroreceptors. Reflex effects on blood pressure and fluid volume in the cat. *Circulation Res.* **42**, 714-20.

Linden, R. J. (1975). Reflexes from the heart. *Prog. cardiovasc. Dis.* **18**, 201-21.

Loewy, A. D. and Burton, H. (1978). Nuclei of the solitary tract: Efferent projections to the lower brain stem and spinal cord of the cat. *J. comp. Neurol.* **181**, 421-50.

McAllen, R. M. and Spyer, K. M. (1976). The location of cardiac vagal preganglionic motoneurones in the medulla of the cat. *J. Physiol.* **258**, 187-204.

——, —— (1978). The baroreceptor input to cardiac vagal motoneurones. *J. Physiol.* **282**, 365-74.

Mancia, G. and Donald, D. E. (1975). Demonstration that the atria, ventricles and lungs each are responsible for a tonic inhibition of the vasomotor center in the dog. *Circulation Res.* **36**, 310-18.

——, Ferrari, A., Gregorini, L., Valentini, R., Ludbrook, J., and Zanchetti, A. (1977). Circulatory reflexes from carotid and extracarotid baroreceptor areas in man. *Circulation Res.* **41**, 309-15.

Manning, J. W. (1977). Intracranial mechanisms of regulation. In *Neural regulation of the heart* (ed. W. C. Randall), pp. 189-209. Oxford University Press, New York.

Mason, J. M. and Ledsome, J. R. (1974). Effects of obstruction of the mitral orifice or distension of the pulmonary vein-atrial junctions on renal and hindlimb vascular resistance in the dog. *Circulation Res.* **35**, 24-32.

Miura, M. and Reis, D. J. (1972). The role of the solitary and paramedian reticular nuclei in mediating cardiovascular reflex responses from carotid baro- and chemoreceptors. *J. Physiol.* 223, 525–48.

Neumayr, R. J., Hare, B. D., and Franz, D. N. (1974). Evidence for bulbospinal control of sympathetic preganglionic neurons by monoaminergic pathways. *Life Sci.* 14, 793–806.

Öberg, B. (1976). Overall cardiovascular regulation. *Ann. Rev. Physiol.* 38, 537–70.

Paintal, A. S. (1969). Mechanisms of stimulation of type J pulmonary receptors. *J. Physiol.* 203, 511–32.

—— (1973). Vagal sensory receptors and their reflex effects. *Physiol. Rev.* 53, 159–227.

Palkovits, M. and Záborszky, L. (1977). Neuroanatomy of central cardiovascular control. Nucleus tractus solitarii: Afferent and efferent neuronal connections in relation to the baroreceptor reflex arc. *Prog. Brain Res.* 47, 9–34.

Pelletier, C. L., Clement, D. L., and Shepherd, J. T. (1972). Comparison of afferent activity of canine aortic and sinus nerves. *Circulation Res.* 31, 557–68.

Peveler, R. C., Bergel, D. H., Gupta, B. N., Sleight, P., and Worley, J. (1980). Modulation of carotid sinus baroreceptor output and carotid sinus mechanical properties by stimulation of efferent sympathetic nerves. In *Arterial baroreceptors and hypertension* (ed. P. Sleight), pp. 6–11. Oxford University Press.

Philippu, A., Demmeler, R., and Roensberg, G. (1974). Effects of centrally-applied drugs on pressor responses to hypothalamic stimulation. *Naunyn-Schmiedebergs Arch. Pharmacol.* 282, 389–400.

Przuntek, H., Guimarães, S., and Philippu, A. (1971). Importance of adrenergic neurons of the brain for the rise of blood pressure evoked by hypothalamic stimulation. *Naunyn-Schmiedebergs Arch. Pharmacol.* 271, 311–19.

Ryall, R. W. (1967). Effect of monoamines upon sympathetic preganglionic neurons. *Circulation Res.* 21, (Suppl. III), 83–7.

Samodelov, L. F., Godehard, E., and Arndt, J. O. (1979). A comparison of the stimulus-response curves of aortic and carotid sinus baroreceptors in decerebrated cats. *Pflügers' Arch.* 383, 47–53.

Schmidt, R. M., Kumada, M., and Sagawa, K. (1972). Cardiovascular responses to various pulsatile pressures in the carotid sinus. *Amer. J. Physiol.* 223, 1–7.

Shepherd, J. T. and Vanhoutte, P. M. (1978). Role of the venous system in circulatory control. *Mayo Clin. Proc.* 53, 247–55.

Sleight, P. and Widdicombe, J. G. (1965). Action potentials in fibres from receptors in the epicardium and myocardium of the dog's left ventricle. *J. Physiol.* 181, 235–58.

Spyer, K. M. (1972). Baroreceptor sensitive neurones in the anterior hypothalamus of the cat. *J. Physiol.* 224, 245–57.

Struyker-Boudier, H., Smeets, G., Brouwer, G., and van Rossum, J.

(1975). Localization of central noradrenergic mechanisms in cardiovascular regulation in rats. *Clin. Sci. Molec. Med.* **48** (Suppl. 2), 277S-278S.

Sybertz, E. J. and Zimmerman, B. G. (1977). Inhibition by guanethidine of chemoreceptor reflex-induced vasodilatation. *Proc. Soc. exp. Biol. Med.* **156**, 426-30.

Takeuchi, T. and Manning, J. W. (1971). Hypothalamic mediation of sinus baroreceptor-evoked muscle cholinergic dilator response. *Amer. J. Physiol.* **224**, 1280-7.

Thames, M. D. and Abboud, F. M. (1979). Reflex inhibition of renal sympathetic nerve activity during myocardial ischaemia mediated by left ventricular receptors with vagal afferents in dogs. *J. clin. Invest.* **63**, 395-402.

Thomas, M. R. and Calaresu, F. R. (1974). Medullary sites involved in hypothalamic inhibition of reflex vagal bradycardia in the cat. *Brain Res.* **80**, 1-16.

Thorén, P. (1976a). Atrial receptors with non-medullated vagal afferents in the cat. *Circulation Res.* **38**, 357-62.

—— (1976b). Activation of left ventricular receptors with non-medullated vagal afferents during occlusion of a coronary artery in the cat. *Amer. J. Cardiol.* **37**, 1046-51.

—— (1979). Role of cardiac vagal C-fibers in cardiovascular control. *Rev. Physiol. Biochem. Pharmac.* **86**, 1-94.

—— and Jones, J. V. (1977). Characteristics of aortic baroreceptor C-fibres in the rabbit. *Acta physiol. Scand.* **99**, 448-56.

Uvnäs, B. (1966). Cholinergic vasodilator nerves. *Fed. Proc.* **25**, 1618-22.

Vatner, S. F. and Braunwald, E. (1975). Cardiovascular control mechanisms in the conscious state. *New Engl. J. Med.* **293**, 970-6.

——, Franklin, D., and Braunwald, E. (1971). Effects of anesthesia and sleep on circulatory response to carotid sinus nerve stimulation. *Amer. J. Physiol.* **220**, 1249-55.

Weiss, G. K. and Crill, W. E. (1969). Carotid sinus nerve: primary afferent depolarization evoked by hypothalamic stimulation. *Brain Res.* **16**, 269-72.

Wing, L. M. H. and Chalmers, J. P. (1974). Participation of central serotonergic neurons in the control of the circulation of the unanesthetized rabbit. — A study using 5,6-dihydroxy-tryptamine in experimental neurogenic and renal hypertension. *Circulation Res.* **35**, 504-13.

Zanchetti, A. (1979). Overview of cardiovascular reflexes in hypertension. *Amer. J. Cardiol.* **44**, 912-18.

Zucker, I. H., Earle, A. M., and Gilmore, J. P. (1979). Changes in the sensitivity of left atrial receptors during reversal of heart failure. *Amer. J. Physiol.* **237**, H555-H559.

3. Regulation of glomerular filtration rate and renal blood-flow

Sodium constitutes nearly 50 per cent of the osmotically active ions in extracellular fluid; since sodium is excluded from cells by the activity of the sodium pump, it is the main determinant of the volume of water in the extracellular compartment. Thus, the control of total body sodium is critical for maintenance of plasma volume, and hence cardiovascular homeostasis.

The renal excretion of sodium is the major factor which determines the mass of sodium in the body, providing that there is an adequate intake. Under normal conditions plasma sodium concentration remains fairly constant due to factors affecting glomerular filtration rate (GFR), renal sodium excretion, and water balance. The present chapter outlines the factors influencing GFR and later chapters will deal with the renal handling of sodium and the control of fluid balance.

Filtration at the glomerulus depends upon the interaction of several forces represented in the equation

$$\text{Net filtration pressure} = P_{GC} - (P_{BC} + \pi_{GC}),$$

where P_{GC} is the glomerular capillary hydrostatic pressure, P_{BC} the Bowman's capsule hydrostatic pressure, and π_{GC} the plasma colloid osmotic pressure. (The colloid osmotic pressure of the filtrate would also be a driving force for filtration, but under normal conditions this variable is negligible, since protein is not filtered through the glomerulus to any great extent.)

If changes in mean arterial blood-pressure were directly transmitted to the glomerular capillaries, they would be expected to alter GFR and salt and water balance. However, the phenomenon of auto-regulation (see p. 4) is particularly prominent in the renal vasculature. Forster and Maes (1947) first showed that, within certain limits, both renal plasma flow and GFR remained relatively constant during experimentally-induced increases in mean arterial blood-pressure. For example, in unanaesthetized rabbits with either intact or denervated kidneys, and with demedullated adrenal glands, a 43 per cent increase in mean arterial blood-pressure caused only a 5 per cent increase in renal plasma flow and an 8 per cent increase in GFR (Forster and Maes 1947). Further work demonstrated that auto-regulation occurred not only when arterial blood-pressure was altered but also when venous pressure was changed (Semple and de Wardener 1959). The observation that auto-regulation of GFR and renal plasma flow was coupled (see also Selkurt, Hall, and Spencer 1949; Shipley and Study 1951) led to the suggestion that changes in afferent rather than efferent arteriolar resistance were responsible; measurements of vascular resistances have confirmed this proposal (Fig. 3.1 and Navar 1978). Subsequent work has been directed towards investigating the mechanisms responsible for such resistance changes and several different hypotheses have been put forward to explain the phenomenon. Of these, two proposals have gained wide support and are outlined here.

3.1. MYOGENIC MECHANISMS

Bayliss (1902) showed that isolated arteries responded to an increase in intraluminal pressure by contraction and to a decrease in pressure by relaxation (see Chapter 1). It was suggested therefore that this 'myogenic' response could account for the changes in renal vascular resistance which occurred during auto-regulation. The time course of renal auto-regulation (within 3 seconds) is consistent with myogenic activity, and it has been shown that agents such as papaverine and procaine (which paralyse smooth-muscle cell contraction), abolish auto-regulation (Thurau and Kramer 1959). However the interpretation of this latter finding is difficult since, regardless of the mechanisms responsible for auto-regulation, the final step must always be smooth-muscle cell contraction. Furthermore, there are some features of renal auto-regulation which are not consistent with the myogenic hypothesis. As mentioned earlier, auto-regulation of GFR and renal plasma flow occurs, not only when arterial blood-pressure is increased, but also when venous pressure is raised. Yet the renal vascular response to an increase in venous pressure is an afferent arteriolar vasodilatation — a response opposite to that which would be expected if a myogenic mechanism was operating. Moreover, increases in venous pressure are accompanied by far greater elevations in interstitial pressure than are comparable increases in arterial pressure, so transmural pressures are quite different under these two conditions, but the magnitudes of the auto-regulatory responses are similar. The role of myogenic mechanisms in renal auto-regulation is, therefore, unclear.

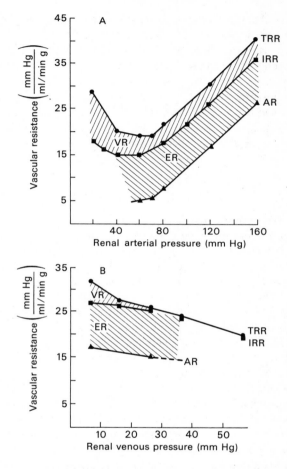

Fig. 3.1. Changes in total renal resistance (TRR), intrarenal resistance (IRR), and preglomerular resistance (AR) in response to changes in renal arterial pressure (A) and renal venous pressure (B). The shaded area between TRR and IRR represents venous resistance (VR); the shaded area between IRR and AR represents efferent resistance (ER). (With permission, Navar (1978).)

3.2. HORMONAL CONTROL

Several workers support the view that an intrarenal, hormonal, feedback mechanism is responsible for the auto-regulation of renal plasma flow and GFR.

The portion of the nephron which marks the transition from the ascending loop of Henle to the distal tubule makes tangential contact with the afferent and efferent arteriole of its own glomerulus; the region of contact is called the macula densa. Collectively, the macula densa, the terminal portion of the afferent arteriole, the efferent arteriole, and the mesangial region between these (the polkissen), are termed the juxtaglomerular apparatus (Fig. 3.2). This anatomical connection between the distal tubule and the vascular supply of the nephron has led workers to suggest that an acute disturbance in distal

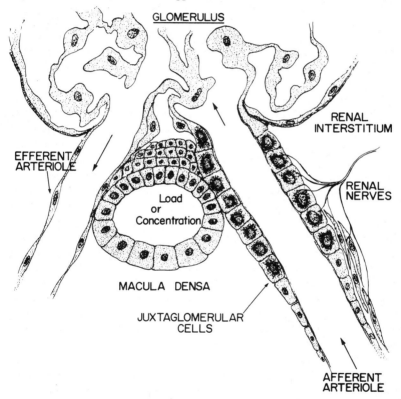

Fig. 3.2. Diagram of the juxtaglomerular apparatus. The two possible receptors, the juxtaglomerular cells of the afferent arteriole and the macula densa, are depicted in their close physical relationship. The renal nerves are depicted as ending on both the juxtaglomerular cells and the smooth muscle cells of the renal afferent arteriole. (With permission, Davis (1971).)

tubular flow might initiate a feedback response which alters renal vascular resistance and thereby affects GFR and renal plasma flow (Guyton, Langston, and Navar 1964; Thurau 1964).

The first evidence for this feedback mechanism was the demonstration, in the rat, that distal tubular microperfusion with isotonic or hypertonic sodium chloride or sodium bromide, caused the proximal tubule to collapse, presumably due to a reduction in GFR (Thurau 1964). For many years it was generally accepted that the sodium concentration at the macula densa was the stimulus for this feedback response. However, in 1970 Schnermann and his co-workers (Schnermann, Wright, Davis, Stackelberg, and Grill 1970) showed that loop perfusion with either sodium sulphate, or a mannitol solution, altered the distal tubular fluid sodium concentration but did not induce a feedback response. Since mannitol and sulphate both reduce the overall reabsorption of salt and water, it was concluded that it was the 'flux' rather than the concentration of sodium which was the stimulus for the feedback response (Schnermann *et al.* 1970). More recently this view has been modified (Schnermann, Ploth, and Hermle 1976). The most reliable method of ensuring that the fluid reaching the macula densa is of known composition is to use retrograde perfusion of the loop from an early distal site. (Previous techniques employed orthograde perfusion from a proximal site which meant that the perfusate travelled through the loop before reaching the macula densa, and thus could have altered in composition.) Using retrograde perfusion, Schnermann *et al.* (1976) demonstrated that a feedback response was obtained only with chloride salts of sodium, potassium, rubidium, caesium, and ammonium, and bromide salts of sodium and potassium; they concluded that it was the halide flux at the macula densa which was the signal for the feedback response. These same workers (Schnermann *et al.* 1976) found that lithium or choline chloride only elicited marked feedback responses when perfused orthogradely. Employing orthograde perfusions, Müller-Suur and Gutsche (1978) showed that there was a feedback response with the chloride salts of sodium, potassium, rubidium, lithium, and choline, but not with sodium sulphate or mannitol (findings consistent with the earlier work (Schnermann *et al.* 1970, 1976)). However, Müller-Suur and Gutsche (1978) also demonstrated a feedback response with sodium acetate, whereas Schnermann *et al.* (1976) did not. The former workers (Müller-Suur and Gutsche 1978) collected and analysed distal tubular fluid during orthograde perfusions to resolve the argument put forward by Schnermann and colleagues regarding the validity of orthograde perfusion. They showed that, during perfusion with sodium or lithium chloride, the sodium and chloride concentrations in the distal tubular fluid rose significantly at the onset of the feedback response whereas during perfusion with sodium acetate only the sodium concentration changed, while perfusion with choline chloride only affected the chloride concentration. Müller-Suur and Gutsche (1978) concluded that there was no single ion species responsible for initiating the feedback response. In another study, using dogs, Bell, Thomas, Williams, and Navar (1978) demonstrated a feedback response during orthograde perfusion with many different isotonic electrolyte solutions, some of which were lacking in sodium, potassium, or chloride. These workers also found that, providing the osmolality of the perfusate was maintained (using mannitol), the sodium chloride concentration could be reduced by as much as 50 per cent without affecting the feedback response; reduction of the osmolality diminished the response (Bell *et al.* 1978). The most remarkable finding in the work of Bell *et al.* (1978) was that isosmotic non-electrolyte solutions containing either mannitol or urea also elicited feedback responses. As an alternative hypothesis they suggested that the total solute delivery, or distal tubular fluid solute concentration, might be important in initiating the feedback response (Bell *et al.* 1978). The results from these different studies are clearly at variance.

Using a different approach, Wright and Persson (1974) showed that the application of electrical currents which made the tubular lumen negative, and thereby enhanced anion efflux, elicited a feedback response, whereas electrically-induced cation efflux did not. This finding supports the hypothesis of Schnermann *et al.* (1976) which favours halide flux as the stimulus for the feedback response. However, under normal physiological conditions, sodium chloride constitutes a major fraction of the solute load presented to the distal tubule, so the same feedback responses would occur irrespective of whether sodium, chloride, or total solute was the stimulus.

The walls of both the afferent and efferent arterioles contain granular cells (Barajas and Latta 1967) which secrete renin (see Chapter 6 and Tobian 1960). Renin is a proteolytic enzyme which cleaves its substrate (an α-2-globulin, synthesized by the liver) to yield a decapeptide— angiotensin I. The latter is acted upon by converting enzyme and gives rise to an octapeptide — angiotensin II. Owing to the proximity of the macula densa to the renin-secreting cells, it has been suggested that a change in the distal tubular fluid delivery alters the release of renin and thus affects the production of angiotensin II, which is a potent vasoconstrictor. The observation that distal tubular microperfusion with sodium chloride caused the proximal tubule to collapse only in normal rats (Thurau 1964) and not in renin-depleted (unilaterally nephrectomized) rats (Thurau and Schnermann 1965) led to the suggestion that increased sodium chloride delivery to the macula densa resulted in renin release, angiotensin II production, and hence afferent arteriolar constriction and a fall in GFR. Later Thurau, Dahlheim, Grüner, Mason, and Granger (1972) demonstrated increased renin activity in the juxtaglomerular apparatus during distal perfusion with sodium chloride, and took this as evidence in favour of their hypothesis. Furthermore, Stowe and Schnermann (1974) found that intravenous infusion of an angiotensin II antagonist impaired feedback control of GFR. However, there is a great deal of evidence which questions the involvement of the renin-angiotensin system in the control of GFR. Firstly, there is an increase in renin release during reductions in renal arterial pressure (Fojas and Schmid 1970; Cowley, Miller, and Guyton 1971), whereas the auto-regulatory response in the vasculature is vasodilatation (see Fig. 3.1). Secondly, there is an inverse relationship between the quantity of sodium chloride delivered to the macula densa and renin release measured in the renal vein (Vander and Miller 1964; Vander and Carlson 1969; DiBona 1971) or renal hilar lymph (Bailie, Loutzenhiser, and Moyer 1972). Thirdly, whereas microperfusion with either sodium chloride or sodium bromide elicits a feedback response (Thurau 1964), only sodium chloride affects renin activity in the juxtaglomerular apparatus (Thurau *et al.* 1972). Finally, several groups of workers (Abe, Kishimoto, and Yamamoto 1976; Anderson, Taher, Cronin, McDonald, and Schrier 1975; Kaloyanides and DiBona 1976) have failed to demonstrate any inhibitory effect of an angiotensin II antagonist on the auto-regulation of renal plasma flow or GFR.

As a result of a recent study, Hall and his colleagues (Hall, Guyton, and Cowley 1977a; Hall, Guyton, Jackson, Coleman, Lohmeier, and Trippodo 1977b) suggested a possible explanation of the discrepancy concerning the role of the renin-angiotensin system in the regulation of renal plasma flow and GFR. Hall *et al.* (1977a) demonstrated that a reduction in renal arterial pressure was associated with auto-regulation of renal plasma flow and GFR in normal dogs; in salt-loaded dogs treated with deoxycorticosterone acetate (in which renin secretion rate was zero), renal plasma flow was well regulated whereas GFR was not. Measurements of arteriolar resistances showed that in normal dogs, the reduction in renal arterial pressure was associated with a fall in afferent arteriolar resistance and a slight increase in efferent arteriolar resistance (similar changes have been reported in the rat (Robertson, Deen, Troy, and Brenner 1972)). However, in renin-depleted dogs, the afferent arteriolar resistance changes were essentially the same as in normal dogs but the efferent vessel resistance fell drastically during the reduction in renal arterial pressure (Hall *et al.* 1977a and Fig. 3.3). Earlier work had

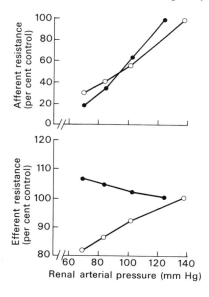

Fig. 3.3. Effect of reducing renal arterial pressure on afferent and efferent arteriolar resistance (calculated by assuming non-equilibrium of filtration pressure) in normal (●, $n = 10$) and renin-depleted (○, $n = 11$) dogs. (With permission, Hall *et al.* (1977a).)

shown that the vasoconstrictor potency of angiotensin II is greater in post- than in pre-glomerular vessels (Regoli and Gauthier 1971; Waugh 1972; Myers, Deen, and Brenner 1975; Fagard, Cowley, Navar, Langford, and Guyton, 1976), and Hall *et al.* (1977a) proposed that GFR is controlled by angiotensin II causing afferent arteriolar vasoconstriction.

Thus during a reduction in renal arterial pressure myogenic mechanisms cause afferent and efferent arteriolar dilatation but in the presence of an intact renin-angiotensin system, the tendency of the efferent vessels to dilate is counteracted by the constrictor action of angiotensin II (Hall *et al.* 1977a). As a further test of this proposal, Hall and his colleagues (Hall *et al.* 1977b) studied the effects of an angiotensin II antagonist on auto-regulation of renal blood-flow and GFR. In dogs maintained on either a normal sodium diet or a low sodium diet a reduction in renal arterial pressure, in the presence of an angiotensin II antagonist, was associated with a fall in efferent arteriolar resistance. In the sodium-replete dogs there was a concomitant increase in renal blood-flow and hence GFR was unchanged whereas in the sodium-depleted animals there was no increase in renal blood-flow and GFR fell. Hall *et al.* (1977b) suggest that the increase in renal blood-flow which occurred in the sodium-replete dogs was due to 'normal dilation of afferent arterioles combined with an excessive dilation of efferent arterioles'. In the sodium-depleted dogs the efferent resistance changes were similar to those which occurred in the normal dogs but the afferent resistance changes were not so marked; the reason for this is unclear. Thus, it appears that the renin-angiotensin system is involved in the control of GFR but the importance of this feedback loop depends on the sodium balance of the animal.

REFERENCES

Abe, Y., Kishimoto, T., and Yamamoto, K. (1976). Effect of angiotensin II antagonist infusion on autoregulation of renal blood flow. *Amer. J. Physiol.* **231**, 1267–71.

Anderson, R. J., Taher, M. S., Cronin, R. E., McDonald, K. M., and Schrier, R. W. (1975). Effect of β-adrenergic blockade and inhibitors of angiotensin II and prostaglandins on renal autoregulation. *Amer. J. Physiol.* **229**, 731–6.

Bailie, M. D., Loutzenhiser, R., and Moyer, S. (1972). Relation of renal hemodynamics to angiotensin II in renal hilar lymph of the dog. *Amer. J. Physiol.* **222**, 1075–8.

Barajas, L. and Latta, H. (1967). Structure of the juxtaglomerular apparatus. *Circulation Res.* **21** (Suppl. II), 15–27.

Bayliss, W. M. (1902). On the local reactions of the arterial wall to changes in internal pressure. *J. Physiol.* **28**, 220–31.

Bell, P. D., Thomas, C., Williams, R. H., and Navar, L. G. (1978). Filtration rate and stop-flow pressure feedback responses to nephron perfusion in the dog. *Amer. J. Physiol.* **234**, F154–F165.

References

Cowley, A. W. Jr., Miller, J. P., and Guyton, A. C. (1971). Open-loop analysis of the renin angiotensin system in the dog. *Circulation Res.* 28, 568–81.

Davis, J.O. (1971). What signals the kidney to release renin? *Circulation Res.* 28, 301–6.

DiBona, G. F. (1971). Effect of mannitol diuresis and ureteral occlusion on distal tubular reabsorption. *Amer. J. Physiol.* 221, 511–14.

Fagard, R. H., Cowley, A. W., Navar, L. G., Langford, H. G., and Guyton, A. C. (1976). Renal responses to slight elevations of renal arterial plasma angiotensin concentration in dogs. *Clin. exp. Pharmacol. Physiol.* 3, 531–8.

Fojas, J. E. and Schmid, H. E. (1970). Renin release, renal auto-regulation, and sodium excretion in the dog. *Amer. J. Physiol.* 219, 464–8.

Forster, R. P. and Maes, J. P. (1947). Effect of experimental neurogenic hypertension on renal blood flow and glomerular filtration rates in intact denervated kidneys of unanesthetized rabbits with adrenal glands demedullated. *Amer. J. Physiol.* 150, 534–40.

Guyton, A. C., Langston, J. B., and Navar, G. (1964). Theory for renal autoregulation by feedback at the juxtaglomerular apparatus. *Circulation Res.* 15, (Suppl. I), 187–96.

Hall, J. E., Guyton, A. C., and Cowley, A. W. Jr. (1977a). Dissociation of renal blood flow and filtration rate autoregulation by renin depletion. *Amer. J. Physiol.* 232, F215–F221.

——, ——, Jackson, T. E., Coleman, T. G., Lohmeier, T. E., and Trippodo, N. C. (1977b). Control of glomerular filtration rate by renin–angiotensin system. *Amer. J. Physiol.* 233, F366–F372.

Kaloyanides, G. J. and DiBona, G. F. (1976). Effect of an angiotensin II antagonist on autoregulation in the isolated dog kidney. *Amer. J. Physiol.* 230, 1078–83.

Müller-Suur, R. and Gutsche, H.-U. (1978). Effect of intratubular substitution of Na^+ and Cl^- ions on the operation of the tubuloglomerular feedback. *Acta physiol. Scand.* 103, 353–62.

Myers, B. D., Deen, W. M., and Brenner, B. M. (1975). Effects of norepinephrine and angiotensin II on the determinants of glomerular ultrafiltration and proximal tubule fluid reabsorption in the rat. *Circulation Res.* 37, 101–10.

Navar, L. G. (1978). Renal autoregulation: perspectives from whole kidney and single nephron studies. *Amer. J. Physiol.* 234, F357–F370.

Regoli, D. and Gauthier, R. (1971). Site of action of angiotensin and other vasoconstrictors on the kidney. *Can. J. Physiol. Pharmacol.* 49, 608–12.

Robertson, C. R., Deen, W. M., Troy, J. L., and Brenner, B. M. (1972). Dynamics of glomerular ultrafiltration in the rat. III. Hemodynamics and autoregulation. *Amer. J. Physiol.* 223, 1191–200.

Schnermann, J., Ploth, D. W., and Hermle, M. (1976). Activation of

tubulo-glomerular feedback by chloride transport. *Pflügers' Arch.* **362**, 229–40.
——, Wright, F. S., Davis, J. M., Stackelberg, W.v., and Grill, G. (1970). Regulation of superficial nephron filtration rate by tubuloglomerular feedback. *Pflügers' Arch.* **318**, 147–75.
Selkurt, E. E., Hall, P. W., and Spencer, M. P. (1949). Influence of graded arterial pressure decrement on renal clearance of creatinine, p-aminohippurate and sodium. *Amer. J. Physiol.* **159**, 369–78.
Semple, S. J. G. and Wardener, H. E. de (1959). Effect of increased renal venous pressure on circulatory "autoregulation" of isolated dog kidney. *Circulation Res.* **7**, 643–8.
Shipley, R. E. and Study, R. S. (1951). Changes in renal blood flow, extraction of inulin, glomerular filtration rate, tissue pressure, and urine flow with acute alterations in renal artery blood pressure. *Amer. J. Physiol.* **167**, 676–88.
Stowe, N. T. and Schnermann, J. (1974). Renin-angiotensin mediation of tubulo-glomerular feedback control of filtration rate. *Fed. Proc.* **33**, 347 (abst. 804).
Thurau, K. (1964). Renal hemodynamics. *Amer. J. Med.* **36**, 698–719.
——, Dahlheim, H., Grüner, A., Mason, J., and Granger, P. (1972). Activation of renin in the single juxtaglomerular apparatus by sodium chloride in the tubular fluid at the macula densa. *Circulation Res.* **30**, (Suppl. II), 182–6.
—— and Kramer, K. (1959). Weitere Untersuchungen zur myogenen Natur der Autoregulation des Nierenkreislaufes. *Pflügers Arch. ges. Physiol.* **269**, 77.
—— and Schnermann, J. (1965). Die Natriumkonzentration an den Macula densa-Zellen als regulierender Faktor für das glomerulumfiltrat. (Mikropunktionsversuche). *Klin. Wschr.* **43**, 410–13.
Tobian, L. (1960). Interrelationship of electrolytes, juxtaglomerular cells and hypertension. *Physiol. Rev.* **40**, 280–312.
Vander, A. J. and Carlson, J. (1969). Mechanism of the effects of furosemide on renin secretion in anesthetized dogs. *Circulation Res.* **25**, 145–52.
—— and Miller, R. (1964). Control of renin secretion in the anesthetized dog. *Amer. J. Physiol.* **207**, 537–46.
Waugh, W. H. (1972). Angiotensin II; local renal effects of physiological increments in concentration. *Can. J. Physiol. Pharmacol.* **50**, 711–16.
Wright, F. S. and Persson, A. E. G. (1974). Effect of changes in distal transepithelial potential difference on feedback control of filtration rate. *Kidney Int.* **6**, 1114A.

4. Proximal tubular reabsorption and glomerulo-tubular balance

4.1. SODIUM REABSORPTION

In man, approximately 170 litres of fluid are filtered by the kidneys daily, of which 150 litres are reabsorbed by the proximal tubules (Knox and Davis 1974). It is generally agreed that an active (energy-requiring) process drives proximal tubular reabsorption. Evidence for this comes from the following observations:

(a) Intravenous infusion of the poorly reabsorbed non-electrolyte, mannitol, causes a marked osmotic diuresis, during which water reabsorption is hindered to a far greater extent than sodium, chloride or bicarbonate, reabsorption (Wesson and Anslow 1948), indicating that ion transport occurs against a concentration gradient.
(b) The cardiac glycoside ouabain, which inhibits a sodium–potassium-activated ATPase, prevents fluid reabsorption in isolated perfused proximal tubules (Burg 1976 and Fig. 4.1).
(c) There is a direct relationship between renal oxygen consumption and absolute sodium reabsorption (Kramer and Deetjen 1960 and Fig. 4.2).

There is evidence to suggest that sodium is the principal ion involved in the active transport. Replacement of sodium with other cations (e.g. lithium, choline) in the fluid bathing and perfusing isolated proximal convoluted tubules inhibits fluid reabsorption completely, whereas omission of chloride has no effect, and omission of bicarbonate only slightly inhibits reabsorption (Burg 1976 and Fig. 4.3).

It was previously thought that the process of fluid reabsorption was uniform along the entire length of the proximal tubule and involved active sodium transport with the subsequent passive movement of chloride and water. More recent evidence, however, has demonstrated that this is not the case. Kokko (1973) showed that the transmembrane potential difference across the proximal convoluted tubule of rabbits varied along the length of the tubule, being negative in the early segments and positive in the later segments (Fig. 4.4). Removal of glucose and amino acids from the perfusate caused the negative potential difference to change to zero (Kokko 1973) and also hindered reabsorption (Burg 1976). It was therefore proposed that in the early segments

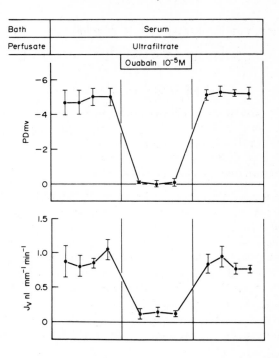

Fig. 4.1. Effect of ouabain on the rate of fluid absorption (J_V) and transepithelial voltage (PD) across isolated perfused rabbit proximal convoluted tubules. Ouabain caused the fluid absorption and voltage to decrease to almost zero. The drug is believed to inhibit active sodium transport which is responsible for the fluid absorption and the voltage. (With permission, Burg (1976).)

of the proximal tubule the negative potential difference was generated by active sodium reabsorption which was coupled to glucose and amino acid transport (Kokko 1973; Barratt, Rector, Kokko, and Seldin 1974). Enhanced sodium transport due to an interaction with organic solutes also occurs in the small intestine and is believed to involve the formation of a complex at the brush border (Schultz, Frizzell, and Nellans 1974; Ullrich 1979).

In the early portion of the proximal tubule, bicarbonate is the anion reabsorbed in preference to chloride. Hence as fluid progresses along the tubule the bicarbonate concentration falls whereas the chloride concentration rises (Weinstein and Szyjewicz 1976). Kokko (1973) showed that when glucose, amino acids, and bicarbonate were removed from

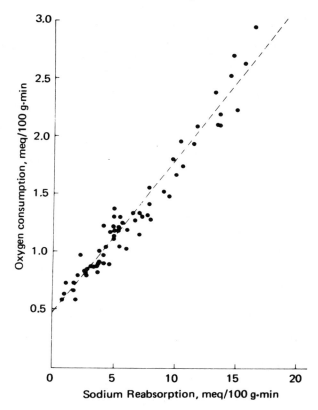

Fig. 4.2. Relationship between oxygen consumption and sodium reabsorption in the dog kidney. Sodium reabsorption was varied by changing the filtered load and by partial blockage of the transport system with drugs. (With permission, Deetjen, Boylan, and Kramer (1975).)

the perfusate of isolated proximal tubules, the potential difference became positive, if chloride was the substituting anion, but not if sulphate was used; in the latter case reabsorption was completely inhibited (Neumann and Rector 1976). Thus the positive potential difference in the later segments of the proximal tubule has been attributed to a chloride diffusion potential (Kokko 1973; Barratt et al. 1974).

To summarize, in the early segments of the proximal convoluted tubule, active sodium transport is enhanced by coupling with organic

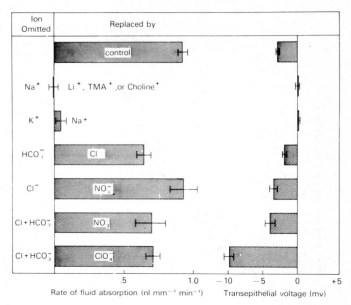

Fig. 4.3. Effect of ion substitutions on the rate of fluid absorption and the transepithelial voltage in isolated rabbit proximal convoluted tubules perfused with an ultra-filtration-like solution and bathed in a serum-like solution. The ion to be investigated was removed entirely from both the perfusate and the bath and was replaced isosmotically with another ion, as indicated. (The exception was K^+, which was removed only from the bath. Removal of K^+ from the lumen only was virtually without effect.) Omission of the cations (Na^+ or K^+) had a great inhibitory effect, because active Na^+ transport (which depends on the presence of K^+) is responsible for the transport processes. Removal of HCO_3^- caused a smaller inhibition. Removal of Cl^- had virtually no effect. (TMA^+, tetramethyl ammonium.) (With permission, Burg (1976).)

solutes and provides the driving force for fluid reabsorption; bicarbonate ions provide the counterbalancing anions and water follows passively. In this segment, approximately 25 per cent of the filtrate is reabsorbed. In the later segments all the organic solutes and bicarbonate have been reabsorbed, and the chloride content of the tubule fluid is high. Active sodium transport therefore occurs at a reduced rate and the major ion movement is the diffusion of chloride down a concentration gradient. Furthermore, since the reflection coefficients of bicarbonate, glucose, and amino acids are higher than that of chloride then, although the total osmolality of the tubule fluid and blood are equal, there is an effective osmotic gradient which drives water out of the

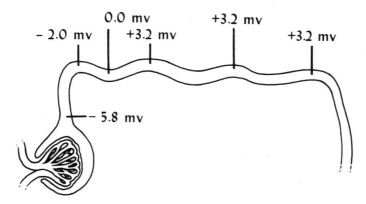

Fig. 4.4. The proposed potential gradient existing in the renal proximal tubule. It is important to note that the depicted PD gradients are the consequence of varying intraluminal constituent concentrations as the fluid courses from the glomerulus to more distal segments of the proximal tubule. The magnitude of the positive PD is principally a function of diffusion potential secondary to the generated high intraluminal chloride concentrations. (With permission, Kokko (1973).)

tubule; sodium and chloride salts follow by solvent drag. The importance of this effective osmotic gradient is indicated by the finding that, in the absence of bicarbonate in the fluid perfusing both the proximal tubular lumen and the peritubular capillaries, sodium and volume flux are reduced by two-thirds. Replacement of the bicarbonate in the peritubular capillaries, but not the luminal fluid, restores the flux (Green and Giebisch 1975). In the later portion of the proximal tubule 35 per cent of the filtrate is reabsorbed of which one-quarter is due to active sodium transport and three-quarters are due to passive chloride movement (Neumann and Rector 1976). In total, therefore, 50 per cent of the proximal tubular fluid reabsorption is attributable to active transport and 50 per cent to passive movements. However, the passive chloride movement depends on the chloride concentration gradient which is established and maintained by the early active transport processes. Hence, the statement at the beginning of this section that an active process drives fluid transport remains valid.

The process of sodium transport from the tubular lumen into the peritubular capillaries involves two steps. Firstly sodium ions *passively* enter the epithelial cells by diffusion and secondly the sodium is *actively* transported from the cells into the lateral intercellular spaces (Fig. 4.5). As a

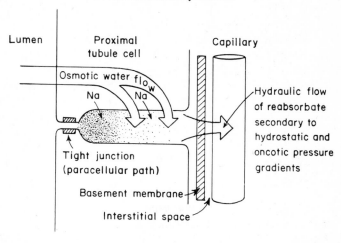

Fig. 4.5. A model for reabsorption of sodium chloride and water by the proximal tubule. Sodium diffuses down an electrochemical concentration gradient into the cell. Sodium is then actively transported into the paracellular space and water follows passively. The oncotic water flow increases the hydrostatic pressure in the space and drives the reabsorbate towards the capillary. Uptake into the peritubular capillaries is accomplished by the balance of Starling forces across the capillary wall. (With permission, Knox and Davis (1974).)

result, the salt concentration (and hence osmolality) in the lateral spaces rises and water moves in by osmosis thereby increasing the hydrostatic pressure in the spaces (Diamond and Tormey 1966). The conductivity across the proximal tubule is too low to permit Starling forces to affect reabsorption (Welling and Grantham 1972). Neither addition of proteins to the luminal fluid of isolated proximal tubules (Imai and Kokko 1974) nor a change in intratubular hydrostatic pressure (Earley and Schrier 1973) affects proximal tubular fluid reabsorption. However, the conductivity across the basement membrane is in the range in which Starling forces could operate (Welling, Welling, and Sullivan 1973) and the fluid moves from the lateral spaces where the hydrostatic pressure is high into the interstitium. The final uptake of fluid into the peritubular capillaries depends, to some extent, on the balance of Starling forces which operate across the capillary walls. Thus

Net pressure for movement into capillaries = $(P_i + \pi_p) - (P_c + \pi_i)$,

where P_i is the interstitial fluid hydrostatic pressure, P_c the peritubular capillary hydrostatic pressure, π_p the peritubular capillary colloid osmotic pressure, and π_i the interstitial fluid colloid osmotic pressure.

Several workers have shown that during extracellular fluid volume expansion with an isotonic saline infusion, there is a fall in proximal tubular sodium reabsorption and, ultimately, natriuresis occurs (Cortney, Mylle, Lassiter, and Gottschalk 1965; Dirks, Cirkensa, and Berliner 1965; Watson 1966; Rector, Sellmann, Martinez-Maldonado, and Seldin 1967). Saline-loading could reduce sodium reabsorption either by reducing active sodium transport or by causing an increased back-flux of fluid into the tubule, due to increased pressure in the lateral intercellular spaces. It is now generally agreed that the natriuresis caused by saline loading is due to increased passive back-diffusion since:

(1) The electrical conductance of the proximal tubular epithelium increases during saline loading (Boulpaep 1972).

(2) In animals in which distal tubular sodium reabsorption is blocked, changes in glomerular filtration rate (GFR) induced by ureteral obstruction are accompanied by parallel changes in proximal sodium reabsorption without any change in oxygen consumption (Lie, Johannesen, and Kiil 1973) — hence passive rather than active sodium transport was being affected.

It seems that during saline-loading the reduction in plasma colloid osmotic pressure, together with the rise in capillary hydrostatic pressure, reduces the movement of fluid from the interstitium into the peritubular capillaries thereby causing a buildup of pressure in the lateral intercellular spaces. As a result the tight junctions between adjacent cells become 'leaky' and fluid fluxes back into the tubule (Fig. 4.6). Brenner, Troy, Daugherty, and McInnes (1973) demonstrated that there was a direct relationship between the peritubular colloid osmotic pressure and proximal tubular fluid reabsorption in the saline-loaded rat, but other workers have failed to show such a relationship in either hydropenic (Rumrich and Ullrich 1968; Conger, Bartoli, and Earley 1976) or saline-expanded (Holzgreve and Schrier 1975) rats. Ott, Haas, Cuche, and Knox (1975) found that the peritubular capillary protein concentration did affect proximal tubular fluid reabsorption in saline-loaded dogs but not in hydropenic animals. They suggested it was only during volume expansion, when the raised capillary hydrostatic pressure had caused the tight junctions to become 'leaky', that the colloid osmotic pressure could exert any effect on fluid reabsorption (Ott et al. 1975). However, unless there is a species-dependent difference, the reason for the discrepancy between the

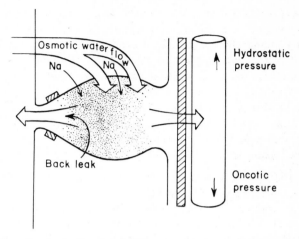

Fig. 4.6. The effects of saline loading on reabsorption of sodium chloride and water by the proximal tubule. Interspace pressure is increased, however, the increased interstitial pressure does not restore reabsorption to control levels, since back-leak into the tubule lumen is markedly augmented. (With permission, Knox and Davis (1974).)

results of Holzgreve and Schrier (1975) in rats and Ott *et al.* (1975) in dogs, is unclear.

4.2. GLOMERULO-TUBULAR BALANCE

A useful index of net tubular reabsorption of salt and water is the ratio of inulin in tubular fluid (TF) compared to plasma (P). Inulin is a polysaccharide which fulfils all the requirements for a suitable marker in renal function tests since:

(1) it is freely filtered at the glomerulus;
(2) it is neither secreted nor reabsorbed by the kidney;
(3) it is neither metabolized nor synthesized by the kidney;
(4) it is non-toxic;
(5) it is conveniently measured.

Any increase in net tubular reabsorption is associated with an increase in TF/P inulin ratio and *vice versa*. Walker, Bott, Oliver, and MacDowell (1941) showed that during experimentally-induced changes in GFR, the TF/P inulin ratio was unchanged at any given point along the proximal tubule. Hence fractional reabsorption must have remained constant. This phenomenon is termed 'glomerulo-tubular balance' (G-T balance) and can be defined as 'events in the proximal tubule in which *absolute* reabsorption changes more or less in parallel with spontaneous or induced changes in filtration rate such that *fractional* reabsorption remains constant' (Tucker and Blantz 1978).

As yet there is no conclusive evidence concerning the mechanisms responsible for G-T balance. One school of thought favours a role of extraluminal events (i.e. peritubular capillary forces) whereas other workers believe that G-T balance is due to intraluminal events in which absolute reabsorption changes as a direct result of a change in filtered load.

4.2.1. Peritubular environment

It has been suggested that the peritubular capillary forces which are, at least in part, responsible for saline-induced natriuresis, may also account for the phenomenon of G-T balance (Lewy and Windhager 1968; Brenner and Troy 1971). Evidence in favour of this comes from the finding of a direct relationship between filtration fraction (GFR divided by renal plasma flow) and proximal tubular reabsorption rate (Lewy and Windhager 1968): since protein is not filtered at the glomerulus, a rise in filtration fraction will cause a parallel increase in peritubular colloid osmotic pressure. As mentioned above, some *in vitro* studies have demonstrated a direct relationship between the protein content of the peritubular fluid and the rate of proximal fluid reabsorption (Spitzer and Windhager 1970; Imai and Kokko 1972). G-T balance occurs during disturbances evoked by renal artery constriction, partial occlusion of the renal vein and partial ureteral obstruction, all of which reduce GFR. In the early experiments of Lewy and Windhager (1968), in which G-T balance was attributed to the influence of peritubular capillary forces, the disturbance imposed was renal vein occlusion. This procedure increases peritubular capillary hydrostatic pressure at the same time as decreasing GFR, and it was suggested that the reduced proximal tubular reabsorption was due to elevated peritubular capillary hydrostatic pressure. However, during aortic or renal artery constriction, peritubular capillary hydrostatic pressure and GFR are both decreased; in that situation the peritubular forces should favour reabsorption, but decreased reabsorption is seen.

4.2.2. Intratubular environment

Burg and Orloff (1968) and Morgan and Berliner (1969) found no significant correlation between proximal tubular load and absolute reabsorption during microperfusion studies in which an artificial perfusate was used. However, in more recent studies, using plasma ultrafiltrate as the perfusion fluid, Bartoli, Conger, and Earley (1973) and Imai, Seldin, and Kokko (1977) showed that proximal tubular reabsorption rate was, to some extent, load-dependent.

In an attempt to clarify the issue, Tucker and Blantz (1978) studied proximal tubular reabsorption in rats under conditions in which filtration rate (load) and interstitial pressure could be varied independently. The results of that study showed a direct correlation between absolute proximal reabsorption and nephron filtration rate, but no correlation between reabsorption and interstitial pressure (Tucker and Blantz 1978). Tucker and Blantz (1978) concluded that intraluminal load does, to some extent, influence proximal tubular reabsorption and could play some part in G-T balance.

4.2.3. Hormonal effects

Howards, Davis, Knox, Wright, and Berliner (1968) infused an albumin solution into dogs to raise the plasma colloid osmotic pressure, and found that proximal tubular sodium reabsorption actually decreased. This seemingly paradoxical situation was subsequently shown by Knox, Schneider, Willis, Strandhoy, Ott, Cuche, Goldsmith, and Arnaud (1974) to be due to hormonal changes. Knox and his co-workers showed that the albumin solution which was used in the earlier experiments (Howards et al. 1968) was very low in calcium and therefore had many sites available for calcium binding. Infusion of this solution decreased the ionized calcium levels in the plasma and hence stimulated parathyroid hormone release (Knox et al. 1974). The phosphaturic effect of parathyroid hormone is mediated by the formation of intracellular cAMP, which activates protein kinase within the renal tubular cell (Aurbach and Heath 1974); such cellular changes could secondarily affect sodium transport. Indeed, Agus, Puschett, Senesky, and Goldberg (1971) and Wen (1974) have shown a fall in proximal tubular sodium reabsorption during administration of physiological doses of parathyroid hormone. These findings do not imply that parathyroid hormone controls sodium reabsorption but they do demonstrate that the hormonal regulation of anion transport can have secondary effects on proximal tubular sodium transport.

It has been suggested that intrarenal production of hormones influences proximal tubular sodium reabsorption and thereby modulates G-T balance. The work of Thurau and his colleagues (outlined in the previous chapter) demonstrates that a change in the solute load of the fluid passing the macula densa can stimulate renin release and thereby lead to intrarenal angiotensin II production. Angiotensin II has been reported to directly inhibit tubular reabsorption (Vander 1963; Leyssac 1965), and Leyssac (1976) incorporated this action of angiotensin II into a new hypothesis to explain the phenomenon of G-T balance. Leyssac (1976) suggested that, rather than a change in GFR leading to changes in sodium reabsorption, the converse was true; thus changes in reabsorption affected intraluminal pressure and hence GFR. There is evidence which indicates that a fall in perfusion pressure or a fall in sodium chloride delivery to the macula densa stimulates renin release and hence angiotensin II production (Chapter 6). According to Leyssac (1976), the two most important intrarenal effects of angiotensin II are vasoconstriction (predominantly in the efferent arterioles (see Chapter 3)), and an inhibition of proximal tubular fluid reabsorption. Thus a fall in filtration pressure reduces proximal intratubular hydrostatic pressure and reduces fluid delivery to the macula densa, thereby stimulating renin release. The subsequent production of angiotensin II increases efferent arteriolar resistance and hence GFR and also inhibits proximal tubular fluid reabsorption. The combined effect of these actions is to increase intraluminal hydrostatic pressure and hence stabilize flow to the distal nephron segments (Fig. 4.7).

However, other workers have shown that angiotensin II stimulates sodium reabsorption (Waugh 1972; Chapman, Riley, and Taton 1977; Hall, Guyton, Trippodo, Lohmeier, McCaa, and Cowley 1977). Thus whether angiotensin II is important in altering tubular sodium reabsorption and if so, by what mechanism, remains to be determined. Poat, Parsons, and Munday (1976) found that whilst low doses of angiotensin II stimulated sodium transport, higher doses inhibited sodium movements in cortical slices. These findings of a dose-related effect could explain some of the discrepant observations mentioned above.

Probably the major effect of angiotensin II on salt and water balance is through its influence on the release of other humoral substances.

74   *Proximal tubular function*

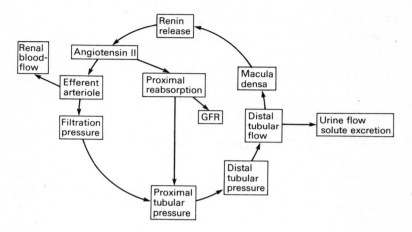

Fig. 4.7. Diagrammatic representation of Leyssac's hypothesis concerning glomerulo-tubular balance. A fall in filtration pressure reduces proximal and distal tubular pressures and flow rates. The reduced fluid delivery to the macula densa elicits renin release; the angiotensin II formed as a result causes efferent arteriolar constriction and inhibits proximal tubular reabsorption, both of which act to increase intraluminal pressure and stabilize flow to the distal nephron. (Modified from Leyssac (1976), with permission.)

Angiotensin II directly inhibits renin release (p. 99), stimulates aldosterone secretion (p. 86), and influences the release of antidiuretic hormone (p. 123); the importance of these hormones in cardiovascular control is discussed elsewhere.

4.2.4. Sympathetic nervous effects

It has been suggested that reabsorption of sodium and water in the proximal tubule may be directly influenced by the renal innervation. Using cross-circulation techniques, Gill and Casper (1969) demonstrated an increase in tubular sodium reabsorption during the increased sympathetic nerve discharge following haemorrhage. They also found that α-adrenoceptor blockade with phenoxybenzamine caused a natriuresis, in the absence of any haemodynamic changes (Gill and Casper 1969). More recently, Bello-Reuss and co-workers (Bello-Reuss, Colindres, Pastoriza-Muñoz, Mueller, and Gottschalk 1975; Bello-Reuss, Trevino, and Gottschalk 1976) demonstrated that unilateral renal denervation in rats increased sodium and water excretion from the denervated kidney, whilst renal nerve stimulation reduced urine volume and sodium excretion without changing GFR or renal plasma flow. In dogs, Zambraski, DiBona, and Kaloyanides (1976a) found that the increase in proximal tubular sodium reabsorption, which occurred during renal nerve stimulation, was blocked by α-adrenoceptor blockade or administration of guanethidine. These workers (Zambraski et al. 1976b) also found that renal tubular sodium reabsorption increased during a reduction of the blood-pressure in the circuit containing the arterial baroreceptors, although the renal perfusion pressure was held constant; the antinatriuresis was not accompanied by any changes in renal haemodynamics and could be reversed by administration of an α-adrenoceptor antagonist or guanethidine. Thus it is possible that a change in sympathetic nerve activity may affect proximal tubular sodium reabsorption and thereby modulate G-T balance.

In summary, under resting conditions, G-T balance operates to minimize changes in sodium excretion due to fluctuations in GFR. However, during extracellular fluid volume expansion it is likely that G-T balance fails, and the resulting natriuresis would tend to reduce the volume towards normal, possibly at the expense of total body sodium. Under conditions such as haemorrhage, G-T balance would not be helpful, since it would act against increasing absolute sodium reabsorption by the nephrons. But it is likely that activation of the sympathetic nervous system and the renin–angiotensin system would override G-T balance in such circumstances.

REFERENCES

Agus, Z. S., Puschett, J. B., Senesky, D., and Goldberg, M. (1971). Mode of action of parathyroid hormone and cyclic adenosine 3′, 5′-monophosphate on renal tubular phosphate reabsorption in the dog. *J. clin. Invest.* **50**, 617–26.

Aurbach, G. D. and Heath, D. A. (1974). Parathyroid hormone and calcitonin regulation of renal function. *Kidney Int.* **6**, 331–45.

Barratt, L. J., Rector, F. C. Jr., Kokko, J. P., and Seldin, D. W. (1974). Factors governing the transepithelial potential difference across the proximal tubule of the rat kidney. *J. clin. Invest.* **53**, 454–64.

Bartoli, E., Conger, J. D., and Earley, L. E. (1973). Effect of intraluminal flow on proximal tubular reabsorption. *J. clin. Invest.* **52**, 843–9.

Bello-Reuss, E., Colindres, R. E., Pastoriza-Muñoz, E., Mueller, R. A., and Gottschalk, C. W. (1975). Effects of acute unilateral renal denervation in the rat. *J. clin. Invest.* **56**, 208–17.

Bello-Reuss, E., Trevino, D. L., and Gottschalk, C. W. (1976). Effect of renal sympathetic nerve stimulation on proximal water and sodium reabsorption. *J. clin. Invest.* **57**, 1104–7.
Boulpaep, E. L. (1972). Permeability changes of the proximal tubule of Necturus during saline loading. *Amer. J. Physiol.* **222**, 517–31.
Brenner, B. and Troy, J. L. (1971). Postglomerular vascular protein concentration: evidence for a causal role in governing fluid reabsorption and glomerulotubular balance by the renal proximal tubule. *J. clin. Invest.* **50**, 336–49.
——, ——, Daugherty, T. M., and McInnes, R. (1973). Quantitative importance of changes in postglomerular colloid osmotic pressure in mediating glomerulotubular balance in the rat. *J. clin. Invest.* **52**, 190–7.
Burg, M. B. (1976). The renal handling of sodium chloride. In *The kidney* (ed. B. M. Brenner and F. C. Rector Jr.), Vol. 1, pp. 272–98. W. B. Saunders, Philadelphia, London and Toronto.
—— and Orloff, J. (1968). Control of fluid reabsorption in the renal proximal tubule. *J. clin. Invest.* **47**, 2016–24.
Chapman, B. J., Riley, A. J., and Taton, A. (1977). Renal effects of angiotensin II in conscious rabbits. *J. Physiol.* **273**, 86P–87P.
Conger, J. D., Bartoli, E., and Earley, L. E. (1976). A study *in vivo* of peritubular oncotic pressure and proximal tubular reabsorption in the rat. *Clin. Sci. Molec. Med.* **51**, 379–92.
Cortney, M. A., Mylle, M., Lassiter, W. E., and Gottschalk, C. W. (1965). Renal tubular transport of water, solute and PAH in rats loaded with isotonic saline. *Amer. J. Physiol.* **209**, 1199–205.
Deetjen, P., Boylan, J. W., and Kramer, K. (1975). *Physiology of kidney and of water balance*, p. 27, Springer Verlag, Heidelberg and New York.
Diamond, J. M. and Tormey, J. McD. (1966). Studies on the structural basis of water transport across epithelial membranes. *Fed. Proc.* **25**, 1458–63.
Dirks, J. H., Cirkensa, W. J., and Berliner, R. W. (1965). Effects of saline infusion on sodium reabsorption by proximal tubule of the dog. *J. clin. Invest.* **44**, 1160–70.
Earley, L. E. and Schrier, R. W. (1973). Intrarenal control of sodium excretion by hemodynamic and physical factors. In *Handbook of physiology*, vol. 8 (ed. J. Orloff and R. W. Berliner), pp. 721–62. American Physiological Society, Washington, DC.
Gill, J. R., Jr. and Casper, A. G. T. (1969). Role of the sympathetic nervous system in the renal response to hemorrhage. *J. clin. Invest.* **48**, 915–22.
Green, R. and Giebisch, G. (1975). Ionic requirements of proximal tubular sodium transport. I. Bicarbonate and chloride. *Amer. J. Physiol.* **229**, 1205–15.
Hall, J. E., Guyton, A. C., Trippodo, N. C., Lohmeier, T. E., McCaa,

R. E., and Cowley, A. W. (1977). Intrarenal control of electrolyte excretion by angiotensin II. *Amer. J. Physiol.* **232**, F538–F544.
Holzgreve, H. and Schrier, R. W. (1975). Variation of proximal tubular reabsorptive capacity by volume expansion and aortic constriction during constancy of peritubular capillary protein concentration in rat kidney. *Pflügers' Arch.* **356**, 73–86.
Howards, S., Davis, B., Knox, F., Wright, F., and Berliner, R. (1968). Depression of fractional sodium reabsorption by the proximal tubule of the dog without sodium diuresis. *J. clin. Invest.* **47**, 1561–72.
Imai, M. and Kokko, J. (1972). Effect of peritubular protein concentration on reabsorption of sodium and water in isolated perfused proximal tubules. *J. clin. Invest.* **51**, 314–25.
——, —— (1974). Transtubular oncotic pressure gradients and net fluid transport in isolated proximal tubules. *Kidney Int.* **6**, 138–45.
——, Seldin, D. W., and Kokko, J. P. (1977). Effect of perfusion rate on the fluxes of water, sodium, chloride and urea across the proximal convoluted tubule. *Kidney Int.* **11**, 18–27.
Knox, F. G. and Davis, B. B. (1974). Role of physical and neuroendocrine factors in proximal electrolyte reabsorption. *Metabolism* **23**, 793–803.
——, Schneider, E. G., Willis, L. R., Strandhoy, J. W., Ott, C. E., Cuche, J.-L., Goldsmith, R. S., and Arnaud, C. D. (1974). Proximal tubule reabsorption after hyperoncotic albumin infusion. *J. clin. Invest.* **53**, 501–7.
Kokko, J. P. (1973). Proximal tubule potential difference. Dependence on glucose, HCO_3 and amino acids. *J. clin. Invest.* **52**, 1362–7.
Kramer, K. and Deetjen, P. (1960). Beziehungen des O_2-Verbrauches der Niere zu Durchblutung und Glomerulumfiltrat bei Änderung des arteriellen Drucks. *Pflügers' Arch. ges. Physiol.* **271**, 782–96.
Lewy, J. E. and Windhager, E. E. (1968). Peritubular control of proximal tubular fluid reabsorption in the rat kidney. *Amer. J. Physiol.* **214**, 943–54.
Leyssac, P. P. (1965). The *in vivo* effect of angiotensin and noradrenaline on the proximal tubular reabsorption of salt in mammalian kidneys. *Acta physiol. Scand.* **64**, 167–75.
—— (1976). The renin–angiotensin system and kidney function. A review of contributions to a new theory. *Acta physiol. Scand.* Suppl. **442**, 1–52.
Lie, M., Johannesen, J., and Kiil, F. (1973). Glomerulotubular balance and renal metabolic rate. *Amer. J. Physiol.* **225**, 1181–6.
Morgan, T. and Berliner, R. W. (1969). An *in vivo* microperfusion study of factors affecting sodium reabsorption in the proximal tubule of the rat kidney. Free communication at the 4th International Congress of Nephrology, Stockholm, p. 232.
Neumann, K. H. and Rector, F. C. (1976). Mechanism of NaCl and water reabsorption in proximal convoluted tubule of rat kidney –

role of chloride concentration gradients. *J. clin. Invest.* **58**, 1110-18.
Ott, C. E., Haas, J. A., Cuche, J.-L., and Knox, F. G. (1975). Effect of increased peritubule protein concentration on proximal tubule reabsorption in the presence and absence of extracellular volume expansion. *J. clin. Invest.* **55**, 612-20.
Poat, J. A., Parsons, B. J., and Munday, K. A. (1976). Effects of angiotensin on transporting epithelia. *J. Endocrinol.* **68**, 2P-3P.
Rector, F. C., Sellmann, J. C., Martinez-Maldonado, M., and Seldin, D. W. (1967). The mechanism of suppression of proximal tubular reabsorption by saline infusions. *J. clin. Invest.* **46**, 47-56.
Rumrich, G. and Ullrich, K. J. (1968). The minimum requirements for the maintenance of sodium chloride reabsorption in the proximal convolution of the mammalian kidney. *J. Physiol.* **197**, 69P-70P.
Schultz, S. G., Frizzell, R. A., and Nellans, H. N. (1974). Ion transport by mammalian small intestine. *Ann. Rev. Physiol.* **36**, 51-91.
Spitzer, A. and Windhager, E. E. (1970). Effect of peritubular oncotic pressure changes on proximal fluid reabsorption. *Amer. J. Physiol.* **218**, 1188-93.
Tucker, B. J. and Blantz, R. C. (1978). Determinants of proximal tubular reabsorption as mechanisms of glomerulotubular balance. *Amer. J. Physiol.* **235**, F142-F150.
Ullrich, K. J. (1979). Sugar, amino acid, and Na^+ cotransport in the proximal tubule. *Ann. Rev. Physiol.* **41**, 181-95.
Vander, A. J. (1963). Inhibition of distal tubular sodium reabsorption by angiotensin II. *Amer. J. Physiol.* **205**, 133-8.
Walker, A. M., Bott, P. A., Oliver, J., and MacDowell, M. C. (1941). The collection and analysis of fluid from single nephrons of the mammalian kidney. *Amer. J. Physiol.* **134**, 580-95.
Watson, J. F. (1966). Effect of saline loading on sodium reabsorption in the dog proximal tubule. *Amer. J. Physiol.* **210**, 781-5.
Waugh, W. H. (1972). Angiotensin II. Local renal effects of physiological increments in concentration. *Can. J. Physiol. Pharmacol.* **50**, 711-16.
Weinstein, S. W. and Szyjewicz, J. (1976). Early postglomerular plasma concentrations of chloride, sodium and inulin in the rat kidney. *Amer. J. Physiol.* **231**, 822-31.
Welling, L. W. and Grantham, J. J. (1972). Physical properties of isolated perfused renal tubules and tubular basement membranes. *J. clin. Invest.* **51**, 1063-75.
Welling, D. J., Welling, L. W., and Sullivan, L. P. (1973). Possible influence of basement membrane on intercellular transport: An analytic formulation. *Fed. Proc.* **32**, 326 (abst. 624).
Wen, S. F. (1974). Micropuncture studies of phosphate transport in the proximal tubule of the dog. The relationship to sodium reabsorption. *J. clin. Invest.* **53**, 143-53.
Wesson, L. G. Jr. and Anslow, W. P. Jr. (1948). Excretion of sodium

and water during osmotic diuresis in the dog. *Amer. J. Physiol.* **153**, 465–74.

Zambraski, E. J., DiBona, G. F., and Kaloyanides, G. J. (1976a). Specificity of neural effect on renal tubular sodium reabsorption. *Proc. Soc. exp. Biol. Med.* **57**, 543–6.

——, ——, —— (1976b). Effect of sympathetic blocking agents on the antinatriuresis of reflex renal nerve stimulation. *J. Pharmacol. exp. Ther.* **198**, 464–72.

5. Distal reabsorption

5.1. LOOP OF HENLE

Salt and water reabsorption in the descending and thin ascending limbs of the loop of Henle is generally believed to be a passive process which depends on the permeability properties of those nephron segments, and on the composition and concentration of the surrounding interstitial fluid. (This process is of major importance in the urine-concentrating mechanism and is discussed in Chapter 9.) Salt transport out of the thick ascending limb of the loop of Henle is an active process. However, in contrast to the proximal tubule, the ion which is transported actively is probably chloride (Burg and Green 1973; Burg and Stoner 1974). The evidence for this comes from work on mammalian and amphibian tubular segments in which the transepithelial potential difference was measured during perfusion with solutions of differing composition. When the principal solute in the perfusate was sodium chloride the luminal voltage was positive, and chloride was transported against an electrochemical gradient. However, when all the chloride was replaced by sulphate or nitrate, the voltage fell to zero, and sodium transport also fell to zero, despite the fact that the membrane permeability to sodium was high. When all the sodium was replaced by choline, the voltage remained positive (Burg and Stoner 1974). From this it was concluded that chloride was actively transported in the thick ascending limb (Burg and Stoner 1974), although the evidence does not rule out the possibility that a co-transport system may operate (Lew, Ferreira, and Moura 1979).

The rate of chloride transport varies directly with the delivery of tubular fluid – a situation analogous to the phenomenon of glomerulo-tubular (G-T) balance in the proximal tubule. However, the reason why the rate of chloride reabsorption is load-dependent is clearer than are the factors responsible for G-T balance in the proximal tubule. In the thick ascending limb of the loop of Henle there is a considerable concentration gradient between the lumen and the interstitium (the lumen being relatively hypotonic). Therefore, an increase in luminal concentration, without any change in active reabsorption, will hinder the process of back-diffusion and hence net absorption will increase (Morgan – reported by de Wardener 1978).

5.2. DISTAL TUBULE

Fluid reaching the distal convoluted tubule is invariably hypotonic, irrespective of whether the final concentration of urine is high or low. The rate of sodium chloride reabsorption in the distal tubule is 25 per cent of the rate of proximal tubular reabsorption. Transepithelial voltage measurements show that in the early portions of the distal tubule the potential difference is positive (i.e. chloride is actively transported) whereas in the later portions the potential difference is negative (i.e. sodium is actively transported; Wright 1971). The permeability of this nephron segment to sodium chloride is lower than that of either the proximal tubule or the loop of Henle and salt reabsorption is under the influence of the hormone aldosterone.

5.2.1. Action of aldosterone

Aldosterone, which is secreted by the outer zone (glomerulosa) of the adrenal cortex, is the most potent, naturally occurring mineralocorticoid. The action of aldosterone on renal electrolyte excretion was first reported by Simpson and Tait (1952) in a study in which the ratio of sodium to potassium in the urine was shown to decrease during administration of aldosterone to adrenalectomized rats. Although the enhancement of sodium reabsorption and potassium secretion predominates in the renal tubules it also occurs in salivary, sweat, gastric, and intestinal epithelia.

The principal renal site of action of aldosterone is generally thought to be the distal tubule. Vander, Malvin, Wild, Lapides, Sullivan, and McMurray (1958) demonstrated a reduction in distal tubular sodium concentration on administration of aldosterone to adrenalectomized rats, whilst there was no detectable change in proximal tubular sodium concentration. Subsequently, Heirholtzer, Wiederholt, Holzgreve, Giebisch, Klose, and Windhager (1965) showed, by micropuncture studies, that the sodium concentration of distal tubular fluid was always greater in adrenalectomized rats than in control rats, and that administration of aldosterone to adrenalectomized rats abolished this difference. Furthermore, Cortney (1969) found that the normal increase in potassium concentration which occurs in the fluid passing along the distal tubule was reduced by adrenalectomy in rats. Although the results of such studies indicate that the site of action of aldosterone is solely on the distal tubule, more recent evidence suggests that aldosterone does have some, albeit small, action on proximal tubular electrolyte transport (Hierholzer and Stolte 1969; Gill, Delea, and Bartter 1972); this remains a contentious point (see discussion by Paillard (1978)).

There is a delay of one to two minutes after aldosterone is administered before the effects on renal electrolyte transport are seen. The sequence of events responsible for this delay are reviewed by Feldman, Funder, and Edelman (1972) and are:

(1) Entry of aldosterone into the cytoplasm by chemical diffusion;
(2) Interaction with a specific receptor system in the cytoplasm;
(3) Translocation of the aldosterone-receptor complex into the nucleus;
(4) Interaction with nuclear chromatin, leading to enhanced RNA transcription, which in turn leads to translation of specific induced proteins;
(5) Mediation of the physiological effects by the induced proteins.

There are three ways in which the specific induced proteins could affect electrolyte transport (see reviews by Fanestil (1969), Feldman *et al.* (1972), Pelletier, Ludens, and Fanestil (1972), and Edelman (1979)); they are illustrated in Fig. 5.1 and involve:

(a) Promoting the passive movement of ions across the luminal (apical) membrane, by means of an increase in membrane permeability (permease theory; Sharp and Leaf 1966; Leaf and MacKnight 1972).
(b) Increasing the inherent activity (or number) of existing sodium pumps on the basal membrane by means of an increase in ATPase activity (pump theory; Goodman, Allen, and Rasmussen 1969).
(c) Stimulating ATP synthesis, and thereby providing energy for the sodium pump (metabolic theory; Edelman 1972).

It is still undecided whether one or all of these mechanisms mediate the effects of aldosterone. Recent evidence suggests that aldosterone could act by both increasing the luminal membrane permeability and by increasing the activity of the sodium pump (Hierholzer and Wiederholt 1976). Interestingly, only the latter effect is inhibited by administration of the protein-synthesis inhibitors actinomycin D or cycloheximide (Hierholzer and Wiederholt 1976), suggesting that the induction of specific proteins is not necessary for an aldosterone-induced increase in luminal membrane permeability.

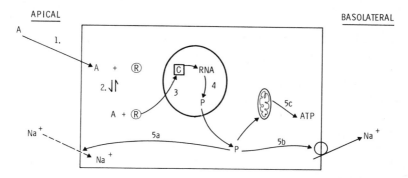

Fig. 5.1. Diagrammatic representation of the mode of action of aldosterone on renal electrolyte transport. 1. Entry of aldosterone (A) into cell; 2. interaction with specific receptor (R); 3. translocation into nucleus and interaction with chromatin (C); 4. RNA transcription and translation of protein (P). Physiological effects of the protein mediated by: 5a, increased membrane permeability; 5b, increased ATPase activity; or 5c, increased ATP synthesis. See text for details.

It was originally thought that aldosterone primarily affected sodium reabsorption and that the concurrent loss of potassium was merely a reflection of an outward movement of cations to maintain electrical neutrality. However, under certain conditions the effects of aldosterone on renal sodium- and potassium-handling are distinct. For example, Malnic, Klose, and Giebisch (1964) showed that infusion of sodium sulphate (a poorly reabsorbed anion) caused a sharp reduction in aldosterone-induced sodium reabsorption whereas potassium secretion was unaffected. Moreover, several workers have observed that administration of protein-synthesis inhibitors prevents the antinatriuretic effect of aldosterone without altering the kaliuresis (Williamson 1963; Fimognari, Fanestil, and Edelman 1967; Wiederholt, Behn, Schoormans, and Hansen 1972). Since protein-synthesis inhibitors prevent the effects of aldosterone on the sodium pump activity, but not the effects on luminal membrane permeability (see above), it is possible that the effects of aldosterone on sodium-handling are mediated by the specific induced proteins whereas the effects on potassium secretion are solely due to the increased membrane permeability. The potassium concentration is much greater in the cell than in the tubular fluid; thus an increase in membrane permeability would permit potassium to move into the tubular lumen down its concentration gradient. This proposal is supported by the findings of Wiederholt et al. (1972). These workers measured the distal tubular transepithelial potential difference with various concentrations of potassium in the luminal perfusate. In normal animals the effect of reducing the potassium concentration of the perfusate 100-fold was to decrease transepithelial potential difference by 52·9 mV; in adrenalectomized rats, such a change in luminal potassium concentration reduced the potential difference by only 32 mV. Administration of aldosterone to the adrenalectomized animals abolished this difference. These changes were not affected by administration of protein-synthesis inhibitors (Wiederholt et al. 1972). In addition to these experimental situations, a further example of a dissociation between the effects of aldosterone on renal sodium- and potassium-handling is seen during the long-term administration of mineralocorticoids to normal man. When deoxycorticosterone is administered to a person on a normal diet there is an initial fall in urinary sodium excretion, an increase in potassium excretion, and an increase in body weight. Potassium excretion is sustained despite the fact that after a few days the antinatriuresis subsides and there is a renal loss of sodium (August, Nelson, and Thorn 1958). This phenomenon has been termed 'escape' and is thought to be a secondary response to extracellular fluid volume expansion, since the time of onset of the escape depends on the rate and intensity of volume expansion (Gross and Möhring 1973). The mechanism of the escape is still not clearly understood, but it may involve the release of an unidentified 'natriuretic' hormone in response to volume expansion (de Wardener, Mills, Clapham, and Hayter 1961; de Wardener 1969, 1977; see later). Alternatively, the influence of the volume expansion on the peritubular capillary Starling forces (see Chapter 4) could hinder proximal tubular sodium reabsorption and thereby cause natriuresis.

5.2.2. Control of aldosterone secretion

Aldosterone secretion is chiefly influenced by the activity of the renin–angiotensin system, the secretion of adrenocorticotrophic hormone (ACTH), and potassium balance (see reviews by Laragh, Sealey, and Brunner 1972; Knochel and White 1973; Fraser, Brown, Lever, Mason, and Robertson 1979).

The observation of an increase in the size of the glomerular zone of the adrenal gland during the development of renal hypertension (Deane and Masson 1951), followed by the demonstration of a direct correlation between the width of the glomerulosa and the granulation of renin-producing (juxtaglomerular) cells in the kidney (Hartroft and Hartroft 1955) led to the suggestion that the renin-angiotensin system stimulated aldosterone release (Gross 1958). Subsequently it was shown that angiotensin II infusion caused an increase in aldosterone secretion (Laragh, Angers, Kelly, and Lieberman 1960; Biron, Koiw, Nowaczynski, Brouillet, and Genest 1961; Ganong, Mulrow, Boryczka, and Cera 1962), the effect being greater when the angiotensin II was administered directly into the adrenal artery than when administered intravenously (Ganong et al. 1962).

Bartter (1956) had suggested earlier that aldosterone production was inversely related to the extracellular fluid volume. It is likely that this relationship is mediated by the renin-angiotensin system, since hypervolaemia suppresses renin secretion and angiotensin II production (Davis, Hartroft, Titus, Carpenter, Ayers, and Spiegel, 1962) and since nephrectomy abolishes the increment in aldosterone secretion normally produced by haemorrhage (Ganong and Mulrow 1962).

Sodium balance also affects aldosterone secretion through changes in renin activity. Dietary sodium restriction increases plasma renin activity and plasma aldosterone, the increase in aldosterone being prevented by administration of either an angiotensin II antagonist or a converting enzyme inhibitor — an agent which prevents the conversion of angiotensin I to angiotensin II (Williams, Hollenberg, Brown, and Mersey 1978). There is also evidence that sodium restriction increases the responsiveness of the adrenal cortex to angiotensin II in dogs (Ganong and Boryczka 1967), rats (Kinson and Singer 1968; Douglas, Hansen, and Catt 1978), and man (Oelkers, Brown, Fraser, Lever, Morton, and Robertson 1974), so for any given dose of angiotensin II aldosterone secretion is greater in the sodium-depleted state. In some hypertensive patients the relationship between sodium balance, renin and aldosterone may be reversed such that there is a reduced adrenal responsiveness to angiotensin II when sodium intake is restricted (Williams, Dluhy, and Moore 1977).

For many years it was accepted that it was the octapeptide angiotensin II which stimulated aldosterone release. However there is now increasing evidence to show that the heptapeptide angiotensin III (the COOH-terminal fragment of angiotensin II) also stimulates aldosterone secretion (Blair-West, Coghlan, Denton, Funder, Scoggins, and Wright 1968; Campbell, Schmitz, and Itskovitz 1977; Davis and Freeman 1977; Bravo 1977, and Fig. 5.2). It has been suggested that angiotensin II and angiotensin III may stimulate the formation of adrenal

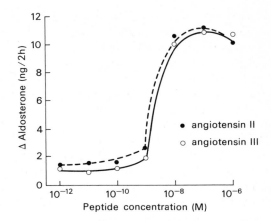

Fig. 5.2. Aldosterone production by isolated adrenal zona glomerulosa cells in response to increasing doses of angiotensin II or angiotensin III. (With permission, Bravo (1977).)

prostaglandins which then augment the peptide-induced aldosterone release (Campbell, Gomez-Sanchez, Adams, Schmitz, and Itskovitz 1979).

Although earlier studies (Cade and Perenich 1965; Marieb and Mulrow 1965) reported that infusions of angiotensin II into rats did not evoke aldosterone release, in those experiments the animals were anaesthetized. More recently Coleman, McCaa, and McCaa (1974) compared the effects of angiotensin II infusion on aldosterone release in conscious and in anaesthetized rats. They demonstrated that the blood-pressure change in response to angiotensin II was similar in both cases, but the increment in aldosterone release was very much less in the anaesthetized animals (Coleman *et al.* 1974). With the present knowledge of the enhancing effects of surgical stress and anaesthesia on ACTH release, it seems likely that in the earlier experiments aldosterone was being maximally secreted under ACTH stimulation before exogenous angiotensin II was administered. Over the short term, the effects of endogenous ACTH secretion on aldosterone release can exceed the effects of angiotensin II (Horton 1973; Espiner, Lun, and Hart 1978) but over a period of days, the effects of ACTH on aldosterone release subside.

Whereas the sodium status of the animal affects aldosterone production indirectly via the renin–angiotensin system, potassium balance can profoundly affect aldosterone secretion directly. In the rat, an increment in the serum potassium concentration of as little as 0.35 m mol/l can increase aldosterone release by as much as 43 per cent (Boyd and Mulrow 1972). This direct stimulatory effect of potassium occurs despite the fact that an increase in serum potassium inhibits renin release (Dluhy, Axelrod, Underwood, and Williams 1972). Potassium balance also alters the responsiveness of the adrenal cortex to various stimuli. Müller (1970) showed that adrenals taken from potassium-restricted rats did not release aldosterone in response to ACTH, angiotensin II, or potassium. Furthermore Boyd, Palmore, and Mulrow (1971) found that potassium was ineffective in causing aldosterone release in rats which had restricted access to sodium and potassium but was effective in sodium-restricted, potassium-repleted animals. The finding that aldosterone release depended on potassium repletion led to the suggestion that an increase in adrenocortical potassium content mediated the effects of the various stimuli on aldosterone release. Baumber, Davis, Johnson, and Witty (1971) showed that angiotensin II, potassium chloride, ACTH, and sodium restriction all increased the potassium content of the adrenals in dogs. However, the way in which this increase in potassium content participates in aldosterone release is unknown.

Aldosterone is the principal antinatriuretic hormone acting in the distal tubule and possibly also in the collecting duct. However, there is some evidence that, under certain conditions, a natriuretic hormone might act to oppose the action of aldosterone. For example, in sodium-loaded rats treated with deoxycorticosterone acetate, blood-volume expansion results in a urinary excretion of sodium almost twice that delivered to the collecting ducts (Sonnenberg, Veress, and Pearce 1971). Moreover, cross-circulation experiments have demonstrated an increase in sodium excretion in the recipient dog during volume expansion of the donor animal; this effect has been attributed to a circulating substance but the nature of this 'natriuretic' hormone has not been determined (see reviews, de Wardener 1977, 1978).

REFERENCES

August, J. T., Nelson, D. H., and Thorn, G. W. (1958). Aldosterone. *New Engl. J. Med.* **259**, 917–23.
Bartter, F. C. (1956). Symposium: Water and electrolytes: role of

aldosterone in normal homeostasis and in certain disease states. *Metabolism* **5**, 369–83.

Baumber, J. S., Davis, J. O., Johnson, J. A., and Witty, R. T. (1971). Increased adrenocortical potassium in association with increased biosynthesis of aldosterone. *Amer. J. Physiol.* **220**, 1094–9.

Biron, P., Koiw, E., Nowaczynski, W., Brouillet, J., and Genest, J. (1961). The effects of intravenous infusions of valine-5-angiotensin II and other pressor agents on urinary electrolytes and corticosteroids, including aldosterone. *J. clin. Invest.* **40**, 338–47.

Blair-West, J. R., Coghlan, J. P., Denton, D. A., Funder, J. W., Scoggins, B. A., and Wright, R. D. (1968). Effects of adrenal steroid withdrawal on chronic renovascular hypertension in adrenalectomized sheep. *Circulation Res.* **23**, 803–9.

Boyd, J. E. and Mulrow, P. J. (1972). Further studies of the influence of potassium upon aldosterone production in the rat. *Endocrinol.* **90**, 299–301.

——, Palmore, W. P., and Mulrow, P. J. (1971). Role of potassium in the control of aldosterone secretion in the rat. *Endocrinol.* **88**, 556–65.

Bravo, E. L. (1977). Regulation of aldosterone secretion: current concepts and newer aspects. *Advan. Nephrol.* **7**, 105–20.

Burg, M. B. and Green, N. (1973). Functions of the thick ascending limb of Henle's loop. *Amer. J. Physiol.* **224**, 659–68.

—— and Stoner, L. (1974). Sodium transport in the distal nephron. *Fed. Proc.* **33**, 31–6.

Cade, R. and Perenich, T. (1965). Secretion of aldosterone by rats. *Amer. J. Physiol.* **208**, 1026–30.

Campbell, W. B., Gomez-Sanchez, C. E., Adams, B. V., Schmitz, J. M., and Itskovitz, H. D. (1979). Attenuation of Angiotensin-II- and III- induced aldosterone release by prostaglandin synthesis inhibitors. *J. clin. Invest.* **64**, 1552–7.

——, Schmitz, J. M., and Itskovitz, H. D. (1977) Adrenal and vascular effects of angiotensin-II and III in sodium depleted rats. *Life Sci.* **20**, 803–10.

Coleman, T. G., McCaa, R. E., and McCaa, C. S. (1974). Effect of angiotensin II on aldosterone secretion in the conscious rat. *J. Endocrinol.* **60**, 421–7.

Cortney, M. A. (1969). Renal tubular transfer of water and electrolytes in adrenalectomized rats. *Amer. J. Physiol.* **216**, 589–98.

Davis, J. O. and Freeman, R. H. (1977). The other angiotensins. *Biochem. Pharmacol.* **26**, 93–7.

——, Hartroft, P. M., Titus, E. O., Carpenter, C. C. J., Ayers, C. R., and Spiegel, H. E. (1962). The role of the renin–angiotensin system in the control of aldosterone secretion. *J. clin. Invest.* **41**, 378–89.

Deane, H. W. and Masson, G. M. C. (1951). Adrenal cortical changes in rats with various types of experimental hypertension. *J. clin. Endocrinol.* **11**, 193–208.

Dluhy, R. G., Axelrod, L., Underwood, R. H., and Williams, G. H. (1972). Studies of the control of plasma aldosterone concentration in normal man. II. Effect of dietary potassium and acute potassium infusion. *J. clin. Invest.* **51**, 1950–7.
Douglas, J., Hansen, J., and Catt, K. J. (1978). Relationships between plasma renin activity and plasma aldosterone in rat after dietary electrolyte changes. *Endocrinol.* **103**, 60–5.
Edelman, I. S. (1972). The initiation mechanism in the action of aldosterone on sodium transport. *J. steroid Biochem.* **3**, 167–71.
—— (1979). Mechanism of action of aldosterone: Energetic and permeability factors. *J. Endocrinol.* **81**, 49P–53P.
Espiner, E. A., Lun, S., and Hart, D. S. (1978). Role of ACTH, angiotensin and potassium in stress-induced aldosterone secretion. *J. steroid Biochem.* **9**, 109–13.
Fanestil, D. D. (1969). Mechanism of action of aldosterone. *Ann. Rev. Med.* **20**, 223–32.
Feldman, D., Funder, J. W., and Edelman, I. S. (1972). Subcellular mechanisms in the action of adrenal steroids. *Amer. J. Med.* **53**, 545–60.
Fimognari, G. M., Fanestil, D. D., and Edelman, I. S. (1967). Induction of RNA and protein synthesis in the action of aldosterone in the rat. *Amer. J. Physiol.* **213**, 954–62.
Fraser, R., Brown, J. J., Lever, A. F., Mason, P. A., and Robertson, J. I. S. (1979). Control of aldosterone secretion. *Clin. Sci.* **56**, 389–99.
Ganong, W. F. and Boryczka, A. T. (1967). Effect of a low sodium diet on aldosterone-stimulating activity of angiotensin II in dogs. *Proc. Soc. exp. Biol. Med.* **124**, 1230–1.
—— and Mulrow, P. J. (1962). Role of the kidney in adrenocortical response to hemorrhage in hypophysectomized dogs. *Endocrinol.* **70**, 182–8.
——, ——, Boryczka, A., and Cera, G. (1962). Evidence for a direct effect of angiotensin II on adrenal cortex of the dog. *Proc. Soc. exp. Biol. Med.* **109**, 381–4.
Gill, J. R. Jr., Delea, C. S., and Bartter, F. C. (1972). A role for sodium-retaining steroids in the regulation of proximal tubular sodium reabsorption in man. *Clin. Sci.* **42**, 423–32.
Goodman, D. B. P., Allen, J. E., and Rasmussen, H. (1969). On the mechanism of action of aldosterone. *Proc. Nat. Acad. Sci. USA* **64**, 330–7.
Gross, F. (1958). Renin und Hypertensin, physiologische oder pathologische Wirkstoffe? *Klin. Wschr.* **36**, 693–706.
—— and Möhring, J. (1973). Renal pharmacology, with special emphasis on aldosterone and angiotensin. *Ann. Rev. Pharmacol.* **13**, 57–90.
Hartroft, P. M. and Hartroft, W. S. (1955). Studies on renal juxtaglomerular cells. II. Correlation of the degree of granulation of

juxtaglomerular cells with width of the zona glomerulosa of the adrenal cortex. *J. exp. Med.* **102**, 205-12.

Hierholzer, K. and Stolte, H. (1969). The proximal and distal tubular action of adrenal steroids on Na reabsorption. *Nephron* **6**, 188-204.

——— and Wiederholt, M. (1976). Some aspects of distal tubular solute and water transport. *Kidney Int.* **9**, 198-213.

———, ———, Holzgreve, H., Giebisch, G., Klose, R. M., and Windhager, E. E. (1965). Micropuncture study of renal transtubular concentration gradients of sodium and potassium in adrenalectomized rats. *Pflügers' Arch. ges. Physiol.* **285**, 193-210.

Horton, R. (1973). Aldosterone: review of its physiology and diagnostic aspects of primary aldosteronism. *Metabolism* **22**, 1525-45.

Kinson, G. A. and Singer, B. (1968). Sensitivity to angiotensin and adrenocorticotrophic hormone in the sodium deficient rat. *Endocrinol.* **83**, 1108-16.

Knochel, J. P. and White, M. G. (1973). The role of aldosterone in renal physiology. *Archs intern. Med.* **131**, 876-84.

Laragh, J. H., Angers, M., Kelly, W. G., and Lieberman, S. (1960). Hypotensive agents and pressor substances. The effect of epinephrine, norepinephrine, angiotensin II, and others on the secretory rate of aldosterone in man. *J. Amer. med. Assoc.* **174**, 234-40.

———, Sealey, J., and Brunner, H. R. (1972). The control of aldosterone secretion in normal and hypertensive man: abnormal renin-aldosterone patterns in low renin hypertension. *Amer. J. Med.* **53**, 649-63.

Leaf, A. and MacKnight, A. D. C. (1972). The site of the aldosterone induced stimulation of sodium transport. *J. steroid. Biochem.* **3**, 237-45.

Lew, V. L., Ferreira, H. G., and Moura, T. (1979). The behaviour of transporting epithelial cells. I. Computer analysis of a basic model. *Proc. roy. Soc. B* **206**, 53-83.

Malnic, G., Klose, R. M., and Giebisch, G. (1964). Micropuncture study of renal potassium excretion in the rat. *Amer. J. Physiol.* **206**, 674-86.

Marieb, N. J. and Mulrow, P. J. (1965). Role of the renin-angiotensin system in the regulation of aldosterone secretion in the rat. *Endocrinol.* **76**, 657-64.

Müller, J. (1970). Steroidogenic effect of stimulators of aldosterone biosynthesis upon separate zones of the rat adrenal cortex: Influence of sodium and potassium deficiency. *Eur. J. clin. Invest.* **1**, 180-7.

Oelkers, W., Brown, J. J., Fraser, R., Lever, A. F., Morton, J. J., and Robertson, J. I. S. (1974). Sensitization of the adrenal cortex to angiotensin II in sodium-deplete man. *Circulation Res.* **34**, 69-77.

Paillard, M. (1977). Effects of aldosterone on renal handling of sodium, potassium and hydrogen ions. *Advan. Nephrol.* **7**, 83-104.

Pelletier, M., Ludens, J. H., and Fanestil, D. D. (1972). The role of aldosterone in active sodium transport. *Archs intern. Med.* **129**, 248-57.

References

Sharp, G. W. G. and Leaf, A. (1966). Mechanism of action of aldosterone. *Physiol. Rev.* **46**, 593-633.

Simpson, S. A. and Tait, J. F. (1952). A quantitative method for the bioassay of the effect of adrenal cortical steroids on mineral metabolism. *Endocrinol.* **50**, 150-61.

Sonnenberg, H., Veress, A. T., and Pearce, J. W. (1972). A humoral component of the natriuretic mechanism in sustained blood volume expansion. *J. clin. Invest.* **51**, 2631-44.

Vander, A. J., Malvin, R. L., Wilde, W. S., Lapides, J., Sullivan, L. P., and McMurray, V. M. (1958). Effects of adrenalectomy and aldosterone on proximal and distal tubular sodium reabsorption. *Proc. Soc. exp. Biol. Med.* **99**, 323-5.

Wardener, H. E. de (1969). Control of sodium reabsorption. *Brit. med. J.* **3**, 611-16 and 676-83.

—— (1977). Editorial review: natriuretic hormone. *Clin. Sci. molec. Med.* **53**, 1-8.

—— (1978). The control of sodium excretion. *Amer. J. Physiol.* **235**, F163-F173.

——, Mills, I. H., Clapham, W. F., and Hayter, C. J. (1961). Studies on the efferent mechanism of the sodium diuresis which follows the administration of intravenous saline in the dog. *Clin. Sci.* **21**, 249-58.

Wiederholt, M., Behn, C., Schoormans, W., and Hansen, L. (1972). Effect of aldosterone on sodium and potassium transport in the kidney. *J. steroid. Biochem.* **3**, 151-9.

Williams, G. H. and Dluhy, R. G. (1972). Aldosterone biosynthesis: Interrelationship of regulatory factors. *Amer. J. Med.* **53**, 595-605.

——, ——, and Moore, T. J. (1977). Aldosterone regulation in essential hypertension. Altered adrenal responsiveness to angiotensin II. *Mayo Clin. Proc.* **52**, 312-16.

——, Hollenberg, N. K., Brown, C., and Mersey, J. H. (1978). Adrenal responses to pharmacological interruption of the renin-angiotensin system in sodium-restricted normal man. *J. clin. Endocrinol. Metab.* **47**, 725-31.

Williamson, H. E. (1963). Mechanism of the antinatriuretic effect of aldosterone. *Biochem. Pharmacol.* **12**, 1449-50.

Wright, F. S. (1971). Increasing magnitude of electrical potential along the renal distal tubule. *Amer. J. Physiol.* **220**, 624-38.

6. Renin release

The previous chapter outlined the effects of the renin-angiotensin system on aldosterone release; below some of the factors which determine the release of renin are considered.

6.1. FACTORS AFFECTING RENIN RELEASE

6.1.1. Baroreceptor hypothesis

Modified myoepithelial cells in the walls of the afferent arterioles (Goormaghtigh 1939; Edelman and Hartroft 1961), and, to a lesser extent, the efferent arterioles (Barajas and Latta 1967) contain granules in which renin is stored. Tobian, Tomboulian, and Janecek (1959) showed that an increase in the perfusion pressure of an isolated kidney preparation reduced the granulation of the renin-secreting (juxtaglomerular) cells, whereas a decrease in pressure caused hypergranulation. These workers suggested that juxtaglomerular-cell granulation was a direct index of secretory activity, and proposed that the juxtaglomerular cells in the walls of the afferent arterioles acted as stretch receptors changing their rate of renin secretion as the distension of the arteriolar wall changed (Tobian *et al.* 1959; Tobian 1960); this is termed the 'baroreceptor hypothesis' for renin release. In the experiments of Tobian and his colleagues, the specific stimulus for renin secretion could not be clearly defined since a number of other variables (such as renal blood-flow, glomerular filtration rate (GFR), and filtered load) would also be expected to change when the perfusion pressure was altered. To clarify this point Blaine and his co-workers (Blaine, Davis, and Witty 1970; Blaine, Davis, and Prewitt 1971) developed a 'non-filtering' kidney model. Filtration was blocked by ureteral clamping and occlusion of the renal artery for two hours to produce renal ischaemia; with filtration blocked, there is no flow in the tubules and hence no change in solute delivery to the macula densa. To remove any other external influences, the contralateral kidney was removed, the remaining kidney was denervated and the animal was adrenalectomized (Blaine *et al.* 1970, 1971). However, in such experiments a reduction in perfusion pressure still evoked a rise in plasma renin activity (Blaine *et al.* 1970, 1971 and Fig. 6.1), which was abolished by papaverine (Witty, Davis, Johnson, and Prewitt 1971);

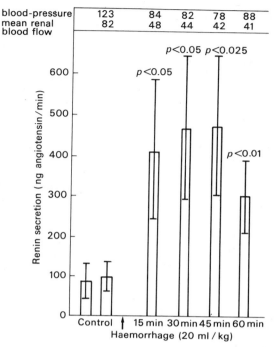

Fig. 6.1. Renin secretion before and after haemorrhage in unilaterally nephrectomized, adrenalectomized dogs in which the remaining kidney was denervated and non-filtering. (Mean arterial blood-pressure (mm Hg); mean renal blood flow (ml/min)). (With permission, Blaine *et al.* (1971).)

this suggests the involvement of some mechanism within the renal vascular tree in the control of renin release. Further support for the baroreceptor hypothesis came from the demonstration by Fray (1976) that in the perfused kidney preparation, renin release was stimulated by either afferent arteriolar constriction or a low perfusion pressure, irrespective of changes in the load of sodium chloride reaching the macula densa. An alternative model for testing the baroreceptor hypothesis involves *in situ* filling of the renal tubules with a low viscosity oil to block filtration and hence fluid delivery to the macula densa (Sadowski and Wocial 1977); under those conditions there is still an inverse relationship between perfusion pressure and renin release.

6.1.2. Macula densa hypothesis

Several workers have shown that renin release can be altered by changing the load, or the composition, of filtrate delivered to the macula densa. There is, at present, disagreement about the nature of the stimulus, acting at the macula densa, which promotes renin release (for a fuller discussion see Chapter 4). Some workers (Thurau, Dahlheim, Grüner, Mason, and Granger 1972; Schnermann, Ploth, and Hermle 1976) showed that an increase in the halide flux across the macula densa increased renin activity in the juxtaglomerular cells whereas others demonstrated an inverse relationship between sodium chloride delivery to the macula densa and renin release into the renal vein (Vander and Miller 1964; Vander and Carlson 1969; DiBona 1971) or renal hilar lymph (Bailie, Loutzenhiser, and Moyer 1972). Although these data might appear conflicting, it should be noted that the various groups of workers measured different indices of juxtaglomerular cell stimulation.

To investigate the effects of distal tubular load on renin release, Humphreys, Reid, Ufferman, and Earley (1971) used a perfused kidney preparation in which the perfusion pressure was held constant whilst distal tubular sodium chloride delivery was altered. They confirmed that there was an inverse relationship between the sodium chloride delivery to the macula densa and renin release (Humphreys *et al.* 1971). Also, Holdsworth, McLean, Morris, Dax, and Johnston (1976) perfused isolated rat glomeruli and showed that a reduction of the sodium concentration of the perfusate from 140 to 110 mmol/l significantly increased the renin-release rate. On balance, the evidence indicates an inverse relationship between sodium chloride delivery to the macula densa and renin release (measured systemically); this is termed the 'macula densa' hypothesis.

Barajas (1971) postulated a common basis for the baroreceptor and macula densa hypotheses using a three-dimensional model of the juxtaglomerular apparatus reconstructed from ultrastructural studies (Fig. 6.2). He found that the efferent arteriole was the only vascular component of the juxtaglomerular apparatus which was in constant contact with the distal tubule; the afferent arteriole made variable contact with the tubule, depending on the degree of distension of either the afferent arteriole or the distal tubule (Barajas 1971). Barajas (1971) suggested that the smaller the contact with the afferent arteriole, the greater the renin release, since the majority of the renin-secreting cells were found

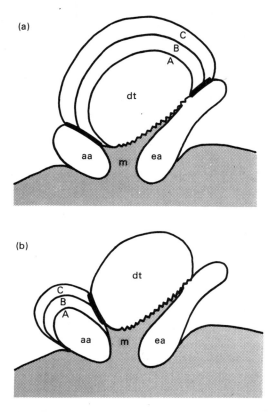

Fig. 6.2. A simplified schematic representation of the proposed functional model of the juxtaglomerular apparatus. The contact between the distal tubule (dt) and the mesangial region (m) and the hilar efferent arteriole (ea) which is interpreted as permanent is represented by wavy lines, whereas the reversible type of contact is represented by heavy lines. (a) As the distal tubule expands (lines B and C) the area of 'reversible' contact with the vascular components increases. (b) Representation of the changes in contact between the distal tubule and the afferent arterioles resulting from changes in the volume of the afferent arteriole.

in parts of the afferent arteriole not in contact with the distal tubule. Thus, during an increase in afferent arteriolar diameter, the area of contact increases and renin secretion falls and vice versa. Alternatively, with an increase in distal tubular fluid load, the tubule is distended, the area of contact increases, and renin secretion falls. However, during a reduction in arterial perfusion pressure, auto-regulatory mechanisms within the renal vasculature cause a vasodilatation to maintain renal blood-flow (see Chapter 3). Such a change would presumably increase the area of contact between the distal tubule and the afferent arteriole which, according to Barajas (1971), should inhibit renin release. Since this is not the case, it has been suggested that it is the auto-regulatory vasodilatation which stimulates renin release when renal perfusion pressure falls (Eide, Løyning, and Kiil 1973; Peart 1978), although the role of vasodilatation in renin release has been questioned (Fray and Karuza 1980).

6.1.3. Plasma sodium concentration

Several groups of workers have shown that plasma sodium concentration influences renin release, but it is still unclear if the effects are due to a direct action on the juxtaglomerular cells, or are mediated by renal baroreceptor or macula densa mechanisms.

Hartroft and Hartroft (1961) observed an inverse relationship between juxtaglomerular cell granulation and plasma sodium concentration and Nash, Rostorfer, Bailie, Wathen, and Schneider (1968) showed that hyponatraemic volume expansion in dogs was accompanied by an increase in renin release which was blocked by intrarenal infusion of hypertonic sodium chloride. Nash *et al.* (1968) suggested that a fall in plasma sodium concentration would give rise to a fall in sodium delivery to the macula densa and thereby stimulate renin release. Shade, Davis, Johnson, and Witty (1972) used the non-filtering kidney model to study the effect of changes in plasma sodium concentration on renin release. They showed that hypertonic sodium chloride infusions into the renal artery caused a marked fall in plasma renin activity only in dogs with intact, filtering kidneys; the non-filtering kidneys failed to respond to changes in plasma sodium concentration (Shade *et al.* 1972). These findings therefore agree with the earlier work of Nash *et al.* (1968) in which it was suggested that the macula densa was involved in the renin response to changes in plasma sodium concentration. However, Newsome and Bartter (1968) found that hypotonic volume expansion in man caused a fall in plasma renin activity despite the reduction in plasma sodium concentration. They also showed that dietary sodium restriction, which reduced the body-weight (and presumably extracellular fluid volume) of their subjects, reduced plasma sodium concentration but increased plasma renin activity (Newsome and Bartter 1968). From this they concluded that there was a closer relationship between plasma renin activity and body fluid balance rather than plasma sodium levels *per se* (Newsome and Bartter 1968). But Brubacher and Vander (1968) found that sodium deprivation consistently elevated plasma renin levels in conscious dogs without changing mean arterial blood-pressure, heart rate, plasma sodium or potassium concentrations, or GFR. Under those conditions neither ganglion blockade nor α- or β-adrenoceptor antagonism altered the renin response, which argues against any participation of the sympathetic nervous system (see later). Moreover, renal baroreceptor or macula densa mechanisms could not have been involved since neither blood-pressure nor GFR changed. Brubacher and Vander (1968) proposed that an unidentified hormone was responsible for renin release during salt restriction. This hormone could act either by altering proximal tubular reabsorption, and hence distal tubular delivery, or by altering the sensitivity of the baroreceptor mechanism, or by directly stimulating the juxtaglomerular cells.

6.1.4. Plasma potassium concentration

Schneider, Davis, Robb, and Baumber (1969) found that potassium chloride infusion caused a decrease in renin release without increasing sodium delivery to the distal tubule, which indicates that potassium may influence renin release directly. This is consistent with the observation that chronic potassium deficiency in the rat causes an increase in renin release independently of any effect on sodium balance (Sealey, Clark, Bull, and Laragh 1970). However, Vander (1970) found that infusion of potassium chloride into the renal arteries caused a natriuresis which was associated with a fall in renin release. Furthermore, Brandis, Keyes, and Windhager (1972) demonstrated that potassium-loading inhibited proximal tubular fluid reabsorption and increased sodium delivery to the distal tubule. Whatever the mechanisms involved, it seems that the macula densa mediates the renin response to changes in plasma potassium concentration, since Shade *et al.* (1972) found that potassium chloride infusion only inhibited renin release in filtering kidneys.

6.1.5. Osmolality

Frederiksen, Leysacc, and Skinner (1975) observed that lowering the osmolality of the fluid superfusing rat renal glomeruli caused an increase in renin-release rate. They correlated the renin-release rate with the reflexion coefficient of the substance used to change the osmolality, and suggested that the juxtaglomerular cells were acting as osmometers and that renin release rate was directly related to juxtaglomerular cell volume.

6.1.6. Sympathetic nerve activity and circulating catecholamines

Both the vascular (Barajas 1964; Doležel, Edvinsson, Owman, and Owman 1976) and tubular (Müller and Barajas 1972) components of the juxtaglomerular apparatus are innervated by sympathetic postganglionic fibres and there is evidence that the activity of the sympathetic nervous system directly affects renin release. Electrical stimulation of the renal nerves (Vander 1965; Johnson, Davis, and Witty 1971; Zambraski and DiBona 1976) or manoeuvres such as bleeding (Buñag, Page, and McCubbin 1966) or tilting (Zanchetti and Stella 1975), which increase sympathetic nerve activity, increase renin secretion independently of vasomotor changes. In addition, intrarenal catecholamine infusion (Johnson *et al.* 1971; Vandongen, Peart, and Boyd 1973) or administration of noradrenalin to the incubation medium of kidney slices (Veyrat and Rosset 1972) stimulates renin release. Conversely suppression of renal nerve activity by hypothalamic stimulation (Zehr and Feigl 1973) or by increased right atrial pressure (Brennan, Malvin, Jochim and Roberts 1971) decreases renin release.

In the isolated, perfused kidney preparation, propranolol prevents renin release caused by sympathetic nerve stimulation (Loeffler, Stockigt, and Ganong 1972) and by isoprenaline, adrenalin, or noradrenalin (Vandongen 1975). In anaesthetized dogs, Almgård and Ljungqvist (1975) found that renin release evoked by exogenous noradrenalin was not dependent upon an intact renal nerve supply, and that propranolol prevented this release of renin without affecting renal blood-flow. A direct effect of propranolol on catecholamine-induced renin release is also indicated by the observation that isoprenaline-stimulated renin release from kidney slices is blocked by propranolol (Capponi and Valloton 1976). Collectively, this evidence suggests that intrarenal β-adrenoceptors are involved in stimulating renin release. But Reid, Schrier, and Earley (1972) found that intrarenal administration of isoprenaline did not cause renin release in anaesthetized dogs, whereas the same dose of isoprenaline given intravenously caused a similar release of renin from intact and denervated kidneys. The intravenous administration of isoprenaline also caused an increase in cardiac output and a fall in total peripheral resistance (Reid *et al.* 1972). It was suggested, therefore, that isoprenaline-induced renin release was mediated by an extrarenal effect of the drug related to the haemodynamic changes which occurred (Reid *et al.* 1972).

There is evidence that, in the whole animal (Pettinger, Keeton, Campbell, and Harper, 1976), and in the isolated kidney (Vandongen and Peart 1974), stimulation of α-adrenoceptors inhibits renin release. For example, Vandongen *et al.* (1973) found that noradrenalin was not a potent stimulus for renin release from the isolated kidney unless given in conjunction with an α-adrenoceptor antagonist; under those conditions the β-adrenoceptor-stimulating properties of noradrenalin were presumably unmasked. Capponi and Valloton (1976) found that noradrenalin inhibited renin release from kidney slices in the absence of an α-adrenoceptor antagonist, but the significance of this finding is obscure since it is generally agreed that release of noradrenalin from renal nerve terminals stimulates renin release.

6.1.7. Hormones

Several circulating hormones can affect renin release. Physiological doses of antidiuretic hormone (ADH) inhibit renin release in anaesthetized (Buñag, Page, and McCubbin 1967) and conscious (Tagawa, Vander, Bonjour, and Malvin 1971) dogs. Furthermore, ADH inhibits renin release in the non-filtering kidney model (Shade, Davis, Johnson, Gotshall, and Spielman 1973); this suggests that ADH may act directly on the juxtaglomerular cells.

Prostaglandins also affect renin release, although the majority of evidence points to an effect of prostaglandins on the renal vasculature rather than directly on juxtaglomerular cells (see review by Lee, 1974). It has been found that prostaglandins produced intrarenally inhibit renin release (Smeby, Sen, and Bumpus 1967). However, there is also evidence that renin release, caused by a fall in perfusion pressure, may require prostaglandins (Berl, Henrich, Erickson, and Schrier 1979; Henrich, Schrier, and Berl 1979).

Angiotensin II can inhibit renin release from kidney slices (Veyrat and Rosset 1972). Furthermore, Shade *et al.* (1973) demonstrated a negative feedback effect of angiotensin II on renin release, in the non-filtering kidney, which was independent of any vascular changes. It is possible that this negative feedback loop may serve to stabilize renin secretion under physiological conditions.

6.2. MECHANISM OF RENIN RELEASE

Since juxtaglomerular cells are modified smooth muscle cells, it is feasible that the processes governing renin release are similar to those involved in smooth muscle excitation − contraction coupling. Excitation of smooth muscle depends on membrane depolarization; Fray (1976), however, proposed that an increase in the stretching of the afferent arteriolar wall led to increased ionic permeability of the juxtaglomerular cell membranes, causing depolarization which *inhibited* renin release, whereas decreased stretching led to hyperpolarization which *stimulated* renin release (Fray 1976). If the mechanisms involved in renin release are not similar to those associated with smooth muscle excitation, then it might be suggested that renin release involves processes similar to those concerned in stimulus−secretion coupling elsewhere (Douglas 1973). In these cases, stimulus−secretion coupling is associated with depolarization and an influx of calcium; however, various workers have shown that renin release is *inhibited* by increased intracellular levels of calcium (Watkins, Davis, Lohmeier, and Freeman 1976; Peart 1977, 1978; Park and Malvin 1978; Fray and Karuza 1980). These results are at variance with those of Chen and Poisner (1976) who reported that increased extracellular calcium levels stimulated renin release from kidney slices when membrane permeability to calcium had been increased. While the details of the mechanisms involved in renin release are elusive at present, there is evidence that the secretion of renin does not involve exocytosis (Baumbach 1980).

REFERENCES

Almgård, L. E. and Ljungqvist, A. (1975). Effect of circulating norepinephrine on the renin release from the denervated kidney. *Scand. J. Urol. Nephrol.* **9**, 125−8.

Bailie, M. D., Loutzenhiser, R., and Moyer, S. (1972). Relation of renal hemodynamics to angiotensin II in renal hilar lymph of the dog. *Amer. J. Physiol.* **222**, 1075−8.

Barajas, L. (1964). The innervation of the juxtaglomerular apparatus. An electron microscopic study of the innervation of the glomerular arterioles. *Lab. Invest.* **13**, 916−29.

—— (1971). Renin secretion: An anatomical basis for tubular control. *Science* **172**, 485−7.

—— and Latta, H. (1967). Structure of the juxtaglomerular apparatus. *Circulation Res.* **21** (Suppl. II), 15−27.

Baumbach, L. (1980). Renin release from isolated rat glomeruli: effects of colchicine, Vinca alkaloids, dimethylsulphoxide, and cytochalasins. *J. Physiol.* **299**, 145-55.

Berl, T., Henrich, W. L., Erickson, A. L., and Schrier, R. W. (1979). Prostaglandins in the beta-adrenergic and baroreceptor-mediated secretion of renin. *Amer. J. Physiol.* **236**, F472-F477.

Blaine, E. H., Davis, J. O., and Prewitt, R. L. (1971). Evidence for a renal vascular receptor in control of renin secretion. *Amer. J. Physiol.* **220**, 1593-7.

——, ——, and Witty, R. T. (1970). Renin release after hemorrhage and after suprarenal aortic constriction in dogs without sodium delivery to the macula densa. *Circulation Res.* **27**, 1081-9.

Brandis, M., Keyes, J., and Windhager, E. E. (1972). Potassium-induced inhibition of proximal tubular fluid reabsorption in rats. *Amer. J. Physiol.* **222**, 421-7.

Brennan, L. A. Jr., Malvin, R. L., Jochim, K. E., and Roberts, D. E. (1971). Influence of right and left atrial receptors on plasma concentrations of ADH and renin. *Amer. J. Physiol.* **221**, 273-8.

Brubacher, E. S. and Vander, A. J. (1968). Sodium deprivation and renin secretion in unanesthetized dogs. *Amer. J. Physiol.* **214**, 15-21.

Buñag, R. D., Page, I. H., and McCubbin, J. W. (1966). Neural stimulation of release of renin. *Circulation Res.* **19**, 851-8.

——, ——, —— (1967). Inhibition of renin release by vasopressin and angiotensin. *Cardiovasc. Res.* **1**, 67-73.

Capponi, A. M. and Vallotton, M. B. (1976). Renin release by rat kidney slices incubated *in vitro*: Role of sodium and of α and β-adrenergic receptors, and effect of vincristine. *Circulation Res.* **39**, 200-3.

Chen, D.-S. and Poisner, A. M. (1976). Direct stimulation of renin release by calcium. *Proc. Soc. exp. Biol. Med.* **152**, 565-7.

DiBona, G. F. (1971). Effect of mannitol diuresis and ureteral occlusion on distal tubular reabsorption. *Amer. J. Physiol.* **221**, 511-14.

Doležel, S., Edvinsson, L., Owman, Ch., and Owman, T. (1976). Fluorescence histochemistry and autoradiography of adrenergic nerves in the renal juxtaglomerular complex of mammals and man, with special regard to the efferent arteriole. *Cell Tissue Res.* **169**, 211-20.

Douglas, W. W. (1973). How do neurones secrete peptides? Exocytosis and its consequences, including 'synaptic vesicle' formation, in the hypothalamo-neurohypophyseal system. *Prog. Brain Res.* **39**, 21-39.

Edelman, R. and Hartroft, P. M. (1961). Localization of renin in juxtaglomerular cells of rabbit and dog through the use of the fluorescent-antibody technique. *Circulation Res.* **9**, 1069-77.

Eide, I., Løyning, E., and Kiil, F. (1973). Evidence for hemodynamic autoregulation of renin release. *Circulation Res.* **32**, 237-45.

Fray, J. C. S. (1976). Stretch receptor model for renin release with evidence from perfused rat kidney. *Amer. J. Physiol.* **231**, 936–44.
——, and Karuza, A. S. (1980). Influence of raising albumin concentration on renin release in isolated perfused rat kidneys. *J. Physiol.* **299**, 45–54.
Frederiksen, O., Leysacc, P. P., and Skinner, S. L. (1975). Sensitive osmometer function of juxtaglomerular cells *in vitro*. *J. Physiol.* **252**, 669–79.
Goormaghtigh, N. (1939). Existence of an endocrine gland in the media of the renal arterioles. *Proc. Soc. exp. Biol. Med.* **42**, 688–9.
Hartroft, W. S. and Hartroft, P. M. (1961). New approaches in the study of cardiovascular disease: aldosterone, renin, hypertension and juxtaglomerular cells. *Fed. Proc.* **20**, 845–54.
Henrich, W. L., Schrier, R. W., and Berl, T. (1979). Mechanisms of renin secretion during hemorrhage in the dog. *J. clin. Invest.* **64**, 1–7.
Holdsworth, S., McLean, A., Morris, B. J., Dax, E., and Johnston, C. I. (1976). Renin release from isolated rat glomeruli. *Clin. Sci. molec. Med.* **51** (Suppl. 3), 97S–99S.
Humphreys, M. H., Reid, I. A., Ufferman, R. C., and Earley, L. E. (1971). Relationship between intrarenal control of sodium reabsorption and renin secretory activity (RSA). *J. clin. Invest.* **50**, 47a.
Johnson, J. A., Davis, J. O., and Witty, R. T. (1971). Effects of catecholamines and renal nerve stimulation on renin release in the nonfiltering kidney. *Circulation Res.* **29**, 646–53.
Lee, J. B. (1974). Prostaglandins and the renal antihypertensive and natriuretic endocrine function. *Recent Prog. Horm. Res.* **30**, 481–522.
Loeffler, J. R., Stockigt, J. R., and Ganong, W. F. (1972). Effect of alpha- and beta-adrenergic blocking agents on the increase in renin secretion produced by stimulation of the renal nerves. *Neuroendocrinol.* **10**, 129–38.
Müller, J. and Barajas, L. (1972). Electron microscopic and histochemical evidence for a tubular innervation in the renal cortex of the monkey. *J. Ultrastruct. Res.* **41**, 533–49.
Nash, F. D., Rostorfer, H. H., Bailie, M. D., Wathen, R. L., and Schneider, E. G. (1968). Renin release: relation to renal sodium load and dissociation from hemodynamic changes. *Circulation Res.* **22**, 473–87.
Newsome, H. H. and Bartter, F. C. (1968). Plasma renin activity in relation to serum sodium concentration and body fluid balance. *J. clin. Endocrinol. Metab.* **28**, 1704–11.
Park, C. S. and Malvin, R. L. (1978). Calcium in the control of renin release. *Amer. J. Physiol.* **235**, F22–F25.
Peart, W. S. (1977). The kidney as an endocrine organ. *Lancet* **II**, 543–7.

Peart, W. S. (1978). Renin release. *Gen. Pharmacol.* **9**, 65–72.
Pettinger, W. A., Keeton, T. K., Campbell, W. B., and Harper, D. C. (1976). Evidence for a renal α-adrenergic receptor inhibiting renin release. *Circulation Res.* **38**, 338–46.
Reid, I. A., Schrier, R. W., and Earley, L. E. (1972). An effect of extrarenal beta adrenergic stimulation on the release of renin. *J. clin. Invest.* **51**, 1861–9.
Sadowski, J. and Wocial, B. (1977). Renin release and autoregulation of blood flow in a new model of non-filtering non-transporting kidney. *J. Physiol.* **266**, 219–33.
Schneider, E. G., Davis, J. O., Robb, C. A., and Baumber, J. S. (1969). Hepatic clearance of renin in canine experimental models for low- and high-output heart failure. *Circulation Res.* **24**, 213–19.
Schnermann, J., Ploth, D. W., and Hermle, M. (1976). Activation of tubulo-glomerular feedback by chloride transport. *Pflügers' Arch.* **362**, 229–40.
Sealey, J. E., Clark, I., Bull, M. B., and Laragh, J. H. (1970). Potassium balance in the control of renin secretion. *J. clin. Invest.* **49**, 2119–27.
Shade, R. E., Davis, J. O., Johnson, J. A., Gotshall, R. W., and Spielman, W. S. (1973). Mechanism of action of angiotensin II and antidiuretic hormone on renin secretion. *Amer. J. Physiol.* **224**, 926–9.
—— , —— , —— , and Witty, R. T. (1972). Effects of renal arterial infusion of sodium and potassium on renin secretion in the dog. *Circulation Res.* **31**, 719–27.
Smeby, R. R., Sen, S., and Bumpus, F. M. (1967). A naturally occurring renin-inhibitor. *Circulation Res.* **21** (Suppl. II), 129–33.
Tagawa, H., Vander, A. J., Bonjour, J.-P., and Malvin, R. L. (1971). Inhibition of renin secretion by vasopressin in unanesthetized sodium deprived dogs. *Amer. J. Physiol.* **220**, 949–51.
Thurau, K. W. C., Dahlheim, H., Grüner, A., Mason, J., and Granger, P. (1972). Activation of renin in the single juxtaglomerular apparatus by sodium chloride in the tubular fluid at the macula densa. *Circulation Res.* **30** (Suppl. II), 182–6.
Tobian, L. (1960). Interrelationship of electrolytes, juxtaglomerular cells and hypertension. *Physiol. Rev.* **40**, 280–312.
—— , Tomboulian, A., and Janecek, J. (1959). The effect of high perfusion pressures on the granulation of juxtaglomerular cells in an isolated kidney. *J. clin. Invest.* **38**, 605–10.
Vander, A. J. (1965). Effect of catecholamines and the renal nerves on renin secretion in anesthetized dogs. *Amer. J. Physiol.* **209**, 659–62.
—— (1970). Direct effects of potassium on renin secretion and renal function. *Amer. J. Physiol.* **219**, 455–9.
—— and Carlson, J. (1969). Mechanism of the effects of furosemide on renin secretion in anesthetized dogs. *Circulation Res.* **25**, 145–52.

Vander, A. J. and Miller, R. (1964). Control of renin secretion in the anesthetized dog. *Amer. J. Physiol.* **207**, 537–46.

Vandongen, R. (1975). Direct intrarenal action of catecholamines on renin secretion. *Clin. exp. Pharmacol. Physiol.* **Suppl. 2**, 103–7.

—— and Peart, W. S. (1974). The inhibition of renin secretion by alpha-adrenergic stimulation of the isolated rat kidney. *Clin. Sci. molec. Med.* **47**, 471–9.

——, ——, and Boyd, G. W. (1973). Adrenergic stimulation of renin secretion in the isolated perfused rat kidney. *Circulation Res.* **32**, 290–6.

Veyrat, R. and Rosset, E. (1972). In vitro renin release by human kidney slices: Effect of norepinephrine, angiotensin II and I and aldosterone. In *Hypertension* (ed. J. Genest and E. Koiw), pp. 45–55. Springer-Verlag, Berlin, Heidelberg, and New York.

Watkins, B. E., Davis, J. O., Lohmeier, T. E., and Freeman, R. H. (1976). Intrarenal site of action of calcium on renin secretion in dogs. *Circulation Res.* **39**, 847–53.

Witty, R. T., Davis, J. O., Johnson, J. A., and Prewitt, R. L. (1971). Effects of papaverine and hemorrhage on renin secretion in the nonfiltering kidney. *Amer. J. Physiol.* **221**, 1666–71.

Zambraski, E. J. and DiBona, G. F. (1976). Angiotensin II in antinatriuresis of low-level renal nerve stimulation. *Amer. J. Physiol.* **231**, 1105–10.

Zanchetti, A. and Stella, A. (1975). Neural control of renin release. *Clin. Sci. molec. Med.* **48**, 215S–223S.

Zehr, J. E. and Feigl, E. O. (1973). Suppression of renin activity by hypothalamic stimulation. *Circulation Res.* **32** (Suppl. I), 17–26.

7. Angiotensin

The physiological effects of renin depend, to a major extent, on the formation of an active polypeptide hormone, angiotensin II. Renin cleaves its substrate — an α-2-globulin formed in the liver — to yield a decapeptide, angiotensin I. The physiological actions of the decapeptide are unclear, but it may stimulate catecholamine secretion from the adrenal medulla (Peach 1971). A converting enzyme, found mainly in the lung capillaries, splits the decapeptide into an octapeptide — angiotensin II; this is the hormone referred to in this chapter as 'angiotensin'. Some of the actions of angiotensin have been discussed elsewhere (p. 73).

7.1. PERIPHERAL EFFECTS ON BLOOD-PRESSURE

Angiotensin is a very powerful vasoconstrictor agent. At low doses this vasoconstrictor action is thought to be mediated by the release of noradrenalin from sympathetic nerve endings (Benelli, Bella, and Gandini 1964; Zimmerman 1967; Hughes and Roth 1969), since the effect is markedly reduced by sympathectomy (Zimmerman 1962) or administration of phentolamine (Goldberg, Joiner, Hyman, and Kadowitz 1975). Angiotensin may also increase vascular tone through increasing noradrenalin biosynthesis from tyrosine (Boadle, Hughes, and Roth 1969) or by increasing adrenal catecholamine output (Braun-Menéndez, Fasciolo, Leloir, and Muñoz, 1940; Peach, Cline, and Watts 1966). At higher doses, however, it may have direct vasoconstrictor effects (McGiff and Fasy 1965). Angiotensin has been detected in the walls of arteries (Swales 1979) and so it is possible that generation of the peptide at this site influences peripheral resistance.

On isolated atria, angiotensin exerts a positive inotropic effect which may be mediated by specific angiotensin-receptor sites (Rioux, Park, and Regoli 1976) since the response is unaffected by α- or β-adrenoceptor antagonists. Freer and his colleagues studied the effects of angiotensin on transmembrane calcium currents in cardiac tissue, and they concluded that the positive inotropic effect of angiotensin was due to its ability to increase ion movement in or through the membrane, probably via 'slow' channels (Freer, Pappano, Peach, Bing, McLean, Vogel, and Sperelakis 1976).

There are drugs which inhibit converting-enzyme activity, and hence angiotensin production, and drugs which antagonize the action of angiotensin. The use of such compounds has permitted an evaluation of the role of the renin–angiotensin system in cardiovascular homeostasis. Several groups of workers (for example, Samuels, Miller, Fray, Haber, and Barger 1973; Gavras, Brunner, Vaughan, and Laragh 1973; Spielman and Davis 1974; Coleman and Guyton 1975; Haber, Sancho, Re, Burton, and Barger 1975; Lohmeier, Cowley, Trippodo, Hall, and Guyton 1977) have shown that, in the conscious or anaesthetized state, inhibitors of the renin–angiotensin system have very little effect on blood-pressure or renal function, when the animals are sodium-replete, but, when dietary sodium intake is restricted, angiotensin antagonists cause a prompt fall in blood-pressure (Fig. 7.1). These results suggest

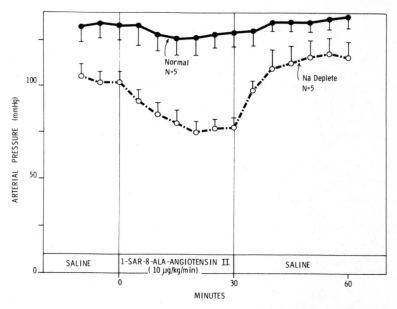

Fig. 7.1. Contrasting responses of mean arterial blood-pressure in normal sodium-replete rats and in sodium-restricted rats infused with 1-Ser-8-Ala-angiotensin II. Bars indicate ± SEM. (With permission, Spielman and Davis (1974).)

that it is only in the sodium-restricted state that the animal depends on the renin–angiotensin system for the maintenance of blood-pressure. However Niarchos, Pickering, Case, Sullivan, and Laragh (1979) found that the administration of an angiotensin antagonist caused a reduction in mean arterial blood-pressure in some sodium-replete subjects but not in others. When sodium intake was restricted, all subjects relied (albeit to varying extents) on the renin–angiotensin system for the maintenance of systemic arterial blood-pressure (Niarchos *et al.* 1979).

Sodium restriction causes a rise in plasma renin activity (although the mechanism behind this is unclear, see Chapter 6) and it has been suggested that the blood-pressure change which occurs in response to angiotensin blockade depends on the circulating plasma renin levels. Indeed, Niarchos *et al.* (1979) found that the sodium-replete subjects who responded to angiotensin antagonism with a fall in blood-pressure, all had higher plasma renin levels than the 'non-responders'. Likewise, Lohmeier *et al.* (1977) reported that, in dogs maintained on a low sodium diet, the largest changes in blood-pressure with angiotensin antagonism occurred in the animals with the highest circulating renin levels. However Brunner and his colleagues (Brunner, Chang, Wallach, Sealey, and Laragh 1972) demonstrated that administration of angiotensin antibodies caused a transient fall in blood-pressure in all states. They suggested (Brunner, Gavras, and Laragh 1974) that blood-pressure in normal and abnormal states is determined by the vasoconstrictor action of angiotensin and by a volume component which depends on sodium balance. Brunner *et al.* (1974) argued that the vasoconstrictor action of angiotensin was of minor importance under normal conditions, but became more significant when sodium intake was restricted. Brunner *et al.* (1974) assumed sodium restriction caused a deficit in extracellular fluid volume, but there is no evidence that this is so. Thus, until the effects of dietary sodium restriction on body fluid and electrolyte balance and renin release are systematically studied, the role of angiotensin in maintaining arterial blood-pressure cannot be defined.

In addition to its effects on angiotensin activity, sodium status can affect other systems which alter cardiovascular control. As mentioned in Chapter 2, the carotid sinus baroreceptors themselves are sensitive to changes in extracellular sodium concentration (Kunze and Brown 1978). Sympathetic nerve function can also be affected by changes in dietary sodium. de Champlain, Krakoff, and Axelrod (1969) found an inverse relationship between dietary sodium intake and the endogenous levels of noradrenalin in the hearts of rats. Animals made hypertensive by deoxycorticosterone implants and salt-loading regained their noradrenalin storage capacity during dietary sodium restriction, and the hypertension also reversed with this procedure; de Champlain *et al.* (1969) suggested that salt loading might disturb some intraneuronal ion balance. Harris (1975) reported that sodium-loading increased the response of the pithed rat preparation to sympathetic nerve stimulation whilst leaving the response to intravenous noradrenalin unchanged; this suggests that sodium-loading increases the release of noradrenalin from sympathetic nerves without any change in effector organ sensitivity. Conversely Raab, Humphreys, Makous, de Grandpré, and Gigee (1952) and Abboud (1974) have demonstrated a decrease in the vasoconstrictor response to noradrenalin during sodium depletion; this phenomenon may also be due to changes in intracellular ion balance.

7.2. CENTRAL EFFECTS ON BLOOD-PRESSURE

In addition to its peripheral actions on vasomotor tone, angiotensin can also evoke a pressor effect by changing central nervous activity. Vertebral artery infusions of angiotensin cause a pressor response (Bickerton and Buckley 1961; Scroop and Lowe 1968, 1969; Gildenberg, Ferrario, and McCubbin 1973), and for several years it was argued that circulating angiotensin might cross the blood-brain barrier and hence exert a central effect. However this postulate is unnecessary (see Phillips 1978), since the pressor response to vertebral artery infusions of angiotensin requires an intact area postrema (Joy and Lowe 1970; Gildenberg *et al.* 1973 and Fig. 7.2), which is a circumventricular organ believed to be lacking in a blood-brain barrier (Dempsey 1973). Morest (1967) demonstrated anatomical connections between the area postrema and the nucleus of the tractus solitarius and the dorsal motor nucleus; these latter regions are thought to relay baroreceptor inputs (see Chapter 2). It is possible, therefore, that circulating angiotensin acts on receptor sites in the area postrema and modulates sympathetic and parasympathetic activity to increase blood-pressure (Dickinson and Yu 1967; Scroop and Lowe 1968, 1969; Ferrario, Gildenberg, and McCubbin 1972; reviewed by Buckley and Jandhyala 1977).

The components of an entire renin-angiotensin system have been demonstrated in the brains of rats and dogs (Fischer-Ferraro, Nahmod, Goldstein, and Finkielman 1971; Ganten, Marquez-Julio, Granger, Hayduk, Karsunky, Boucher, and Genest 1971; but see Reid 1977). Moreover, centrally administered angiotensin elicits a pressor response

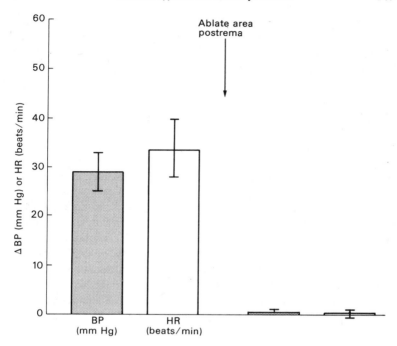

Fig. 7.2. The effects of intravertebral infusions of angiotensin II on mean arterial blood-pressure (BP) and heart rate (HR) before and after ablation of the area postrema. The histograms represent the mean ± SEM for nine dogs. (With permission, Joy and Lowe (1970).)

which involves activation of the sympathetic nervous system and release of antidiuretic hormone (Hoffman and Phillips 1977, see later p. 123). This effect is distinct from that described above, since lesions in the area postrema do not affect the pressor response to centrally administered angiotensin (Gildenberg et al. 1973; Phillips 1978); the site which mediates this pressor effect is in the region of the subnucleus medialis (Buckley, Smookler, Severs, and Deuben 1977). It appears that there is a barrier which prevents angiotensin in the cerebrospinal fluid from reaching receptors available to the blood-borne peptide, since angiotensin applied to the ependymal surface of the area postrema does not affect blood-pressure (Phillips and Hoffman 1977).

There is recent evidence indicating that the adrenal medullary response to haemorrhage may be mediated by the central action of angiotensin (Feuerstein, Boonyaviroj, Gutman, Khosla, and Bumpus 1977). Feuerstein *et al.* (1977) showed that administration of an angiotensin antagonist to cats abolished adrenal medullary catecholamine secretion in response to haemorrhage. Furthermore, they could not demonstrate any catecholamine release from denervated adrenals, even in the presence of angiotensin. They suggested, therefore, that angiotensin was acting centrally to alter sympathetic nerve activity and hence adrenal medullary secretion (Feuerstein *et al.* 1977); whether the angiotensin was produced peripherally or centrally is unclear. Mann, Phillips, Dietz, Haebara, and Ganten (1978) have recently studied the effects of both peripheral and central administration of an angiotensin antagonist (Saralasin) to rats made hypertensive by a variety of manoeuvres. In their control animals neither peripheral nor central administration of Saralasin affected mean arterial blood-pressure. But, in Goldblatt two-kidney hypertensive rats, administration of Saralasin into the cerebral ventricles caused a fall in blood-pressure, while intravenous administration was without effect. Conversely, Saralasin increased blood-pressure in rats made hypertensive by deoxycorticosterone acetate and saline, irrespective of whether the drug was administered centrally .or peripherally. Intravenously administered Saralasin also increased blood-pressure in spontaneously hypertensive rats, but central administration of Saralasin reduced blood-pressure. These results clearly show that angiotensin can influence blood-pressure in certain hypertensive states and that the central and peripheral effects can be quite distinct.

Recently, it has been found that intravenous administration of angiotensin inhibits the baroreflex response to acute elevations in systemic arterial pressure (Ismay, Lumbers, and Stevens 1979). Thus the renin–angiotensin system may be involved in diverse ways in cardiovascular homeostasis. A further central effect of angiotensin — the influence on drinking behaviour — is discussed in Chapter 10.

REFERENCES

Abboud, F. M. (1974). Effects of sodium, angiotensin and steroids on vascular reactivity in man. *Fed. Proc.* 33, 143–9.

Benelli, G., Bella, D. D., and Gandini, A. (1964). Angiotensin and peripheral sympathetic nerve activity. *Brit. J. Pharmacol. Chemother.* 22, 211–19.

Bickerton, R. K. and Buckley, J. P. (1961). Evidence for a central

mechanism in angiotensin induced hypertension. *Proc. Soc. exp. Biol. Med.* **106**, 834–6.
Boadle, M. C., Hughes, J., and Roth, R. H. (1969). Angiotensin accelerates catecholamine biosynthesis in sympathetically innervated tissues. *Nature (Lond.)* **222**, 987–8.
Braun-Menéndez, E., Fasciolo, J. C., Leloir, L. F., and Muñoz, J. M. (1940). The substances causing renal hypertension. *J. Physiol.* **98**, 283–98.
Brunner, H. R., Chang, P., Wallach, R., Sealey, J. E., and Laragh, J. H. (1972). Angiotensin II vascular receptors; their avidity in relationship to sodium balance, the autonomic nervous system, and hypertension. *J. clin. Invest.* **51**, 58–67.
——, Gavras, H., and Laragh, J. H. (1974). Specific inhibition of the renin–angiotensin system: A key to understanding blood pressure regulation. *Prog. cardiovasc. Dis.* **17**, 87–98.
Buckley, J. P. and Jandhyala, B. S. (1977). Central cardiovascular effects of angiotensin. *Life Sci.* **20**, 1485–93.
——, Smookler, H. H., Severs, W. B., and Deuben, R. R. (1977). A central site of action of angiotensin II and its possible role in the regulation of the cardiovascular system. In *Central actions of angiotensin and related hormones* (ed. J. P. Buckley and C. M. Ferrario), pp. 149–55. Pergamon Press, New York.
Champlain, J. de., Krakoff, L. R., and Axelrod, J. (1969). Interrelationship of sodium intake, hypertension and norepinephrine storage in the rat. *Circulation Res.* **24** (Suppl. I), 75–92.
Coleman, T. G. and Guyton, A. C. (1975). The pressor role of angiotensin in salt deprivation and renal hypertension in rats. *Clin. Sci. Molec. Med.* **48**, 45S–48S.
Dempsey, E. W. (1973). Neural and vascular ultrastructure of the area postrema in the rat. *J. comp. Neurol.* **150**, 177–200.
Dickinson, C. J. and Yu, R. (1967). Mechanisms involved in the progressive pressor response to very small amounts of angiotensin in conscious rabbits. *Circulation Res.* **21** (Suppl. II), 157–63.
Ferrario, C. M., Gildenberg, P. L., and McCubbin, J. W. (1972). Cardiovascular effects of angiotensin mediated by the central nervous system. *Circulation Res.* **30**, 257–62.
Feuerstein, G., Boonyaviroj, P., Gutman, Y., Khosla, M. C., and Bumpus, F. M. (1977). Adrenal catecholamine response to haemorrhage abolished by an angiotensin antagonist. *Eur. J. Pharmacol.* **41**, 85–6.
Fischer-Ferraro, C., Nahmod, V. E., Goldstein, D. J., and Finkielman, S. (1971). Angiotensin and renin in rat and dog brain. *J. exp. Med.* **133**, 353–61.
Freer, R. J., Pappano, A. J., Peach, M. J., Bing, K. T., McLean, M. J., Vogel, S., and Sperelakis, N. (1976). Mechanism for the positive inotropic effect of angiotensin II on isolated cardiac muscle. *Circulation Res.* **39**, 178–83.

Ganten, D., Marquez-Julio, A., Granger, P., Hayduk, K., Karsunky, K. P., Boucher, R., and Genest, J. (1971). Renin in dog brain. *Amer. J. Physiol.* **221**, 1733–7.

Gavras, H., Brunner, H. R., Vaughan, E. D., Jr., and Laragh, J. H. (1973). Angiotensin–sodium interaction in blood pressure maintenance of renal hypertensive and normotensive rats. *Science* **180**, 1369–71.

Gildenberg, P. L., Ferrario, C. M., and McCubbin, J. W. (1973). Two sites of cardiovascular action of angiotensin II in the brain of the dog. *Clin. Sci.* **44**, 417–20.

Goldberg, M. R., Joiner, P. D., Hyman, A. L., and Kadowitz, P. J. (1975). Unusual venoconstrictor effects of angiotensin II. *Proc. Soc. exp. Biol. Med.* **149**, 707–13.

Haber, E., Sancho, J., Re, R., Burton, J., and Barger, A. C. (1975). The role of the renin–angiotensin–aldosterone system in cardiovascular homeostasis in normal man. *Clin. Sci. Molec. Med.* **48**, 49s–52s.

Harris, G. S. (1975). Relationship between body sodium and arterial pressure. *Clin. exp. Pharmacol. Physiol.* Suppl. **2**, 123–6.

Hoffman, W. E. and Phillips, M. I. (1977). The role of ADH in pressor response to intraventricular angiotensin II. In *Central actions of angiotensin and related hormones* (ed. J. P. Buckley and C. M. Ferrario), pp. 307–14. Pergamon Press, New York.

Hughes, J. and Roth, R. H. (1969). Enhanced release of transmitter during sympathetic nerve stimulation in the presence of angiotensin. *Brit. J. Pharmacol.* **37**, 516P–517P.

Ismay, M. J. A., Lumbers, E. R., and Stevens, A. D. (1979). The action of angiotensin II on the baroreflex response of the conscious ewe and the conscious foetus. *J. Physiol.* **288**, 467–79.

Joy, M. D. and Lowe, R. D. (1970). The site of cardiovascular action of angiotensin II in the brain. *Clin. Sci.* **39**, 327–36.

Kunze, D. L. and Brown, A. M. (1978). Sodium sensitivity of baroreceptors. Reflex effects on blood pressure and fluid volume in the cat. *Circulation Res.* **42**, 714–20.

Lohmeier, T. E., Cowley, A. W. Jr., Trippodo, N. C., Hall, J. E., and Guyton, A. C. (1977). Effects of endogenous angiotensin II on renal sodium excretion and renal hemodynamics. *Amer. J. Physiol.* **233**, F388–F395.

Mann, J. F. E., Phillips, M. I., Dietz, R., Haebara, H., and Ganten, D. (1978). Effects of central and peripheral angiotensin blockade in hypertensive rats. *Amer. J. Physiol.* **234**, H629–H637.

McGiff, J. C. and Fasy, T. M. (1965). The relationship of the renal vascular activity of angiotensin II to the autonomic nervous sytem. *J. clin. Invest.* **44**, 1911–23.

Morest, D. K. (1967). Experimental study of the projections of the nucleus of the tractus solitarius and the area postrema in the cat. *J. comp. Neurol.* **130**, 277–99.

Niarchos, A. P., Pickering, T. G., Case, D. B., Sullivan, P., and Laragh, J. H. (1979). Role of the renin–angiotensin system in blood pressure regulation. The cardiovascular effects of converting enzyme inhibition in normotensive subjects. *Circulation Res.* **45**, 829–37.

Peach, M. J. (1971). Adrenal medullary stimulation induced by angiotensin I, angiotensin II and analogues. *Circulation Res.* **28-9** (Suppl. II), 107–17.

——, Cline, W. H., Jr., and Watts, D. T. (1966). Release of adrenal catecholamines by angiotensin II. *Circulation Res.* **19**, 571–5.

Phillips, M. I. (1978). Angiotensin in the brain. *Neuroendocrinol.* **25**, 354–77.

—— and Hoffman, W. E. (1977). Sensitive sites in the brain for the blood pressure and drinking responses to angiotensin II. In *Central actions of angiotensin and related hormones* (ed. J. P. Buckley and C. M. Ferrario), pp. 325–56. Pergamon Press, New York.

Raab, W., Humphreys, R. J., Makous, N., de Grandpré, R., and Gigee, W. (1952). Pressor effects of epinephrine, norepinephrine and desoxycorticosterone acetate (DCA) weakened by sodium withdrawal. *Circulation* **6**, 373–7.

Reid, I. A. (1977). Is there a brain renin–angiotensin system? *Circulation Res.* **41**, 147–53.

Rioux, F., Park, W. K., and Regoli, D. (1976). Characterization of angiotensin receptors in rabbit isolated atria. *Can. J. Physiol. Pharmacol.* **54**, 229–37.

Samuels, A. I., Miller, E. D., Fray, C. S., Haber, E., and Barger, A. C. (1973). The regulation of pressure by the renin–angiotensin system. *Fed. Proc.* **32**, 380.

Scroop, G. C. and Lowe, R. D. (1968). Central pressor effect of angiotensin mediated by the parasympathetic nervous system. *Nature (Lond.)* **220**, 1331–2.

——, —— (1969). Efferent pathways of the cardiovascular response to vertebral artery infusions of angiotensin in the dog. *Clin. Sci.* **37**, 605–19.

Spielman, W. S. and Davis, J. O. (1974). The renin–angiotensin system and aldosterone secretion during sodium depletion in the rat. *Circulation Res.* **35**, 615–24.

Swales, J. D. (1979). Arterial wall or plasma renin in hypertension. *Clin. Sci.* **56**, 293–8.

Zimmerman, B. G. (1962). Effect of acute sympathectomy on responses to angiotensin and norepinephrine. *Circulation Res.* **11**, 780–7.

—— (1967). Evaluation of peripheral and central components of action of angiotensin on the sympathetic nervous system. *J. Pharmacol. exp. Ther.* **158**, 1–10.

8. Antidiuretic hormone

The renal response to a reduction in extracellular fluid volume is an increase in sodium reabsorption. In the proximal tubule, the sodium reabsorbed is accompanied by water, and this tends to maintain extracellular fluid volume and osmolality. In the distal nephron, however, sodium reabsorption is not necessarily accompanied by water movements, and addition of this sodium to the extracellular fluid increases its osmolality. Since sodium is not free to move into cells, it causes an efflux of intracellular water which increases extracellular fluid volume and minimizes the increase in osmolality. However, this water shift cannot return the osmolality of the extracellular fluid to normal since extra osmoles have been added to the system; the fine control of extracellular osmolality is achieved by adjustments in water balance (Gauer 1968; Young, Pan, and Guyton 1977).

Changes in extracellular fluid osmolality and volume influence the renin–angiotensin–aldosterone system (see Chapter 6) and release of antidiuretic hormone (see below), so it is wrong to consider that osmolality and volume can be controlled independently. However, the osmolality of extracellular fluid is largely determined by thirst mechanisms which control fluid intake, (see Chapter 10) and by antidiuretic hormone which regulates free water clearance. This hormone also has other actions important in cardiovascular homeostasis (see below).

8.1. BIOSYNTHESIS OF ANTIDIURETIC HORMONE

The active antidiuretic hormone (ADH) in most mammalian species is an octapeptide—arginine vasopressin. Sachs and his colleagues (Takabatake and Sachs 1964; Sachs 1967, 1969; Sachs, Fawcett, Takabatake, and Portanova 1969) have shown that a biologically inactive protein is synthesized in the supraoptic and paraventricular nuclei of the anterior hypothalamus. There it is cleaved into the biologically active octapeptide, and incorporated into membrane-bound neurosecretory granules. The ADH activity which they measured in the hypothalamus was several times greater than in the posterior pituitary. Moreover, whereas the hypothalamic–median eminence complex was capable of synthesizing ADH *de novo,* the posterior pituitary was not (Takabatake and Sachs 1964) and section of the supraoptico-hypophyseal tract prevented the appearance of any ADH in the posterior pituitary. Sachs et al. (1969) concluded that ADH is synthesized in the supraoptic and paraventricular nuclei of the hypothalamus and transported down the supraoptico-hypophyseal tract to the posterior pituitary for storage and release. The above-mentioned hypothalamic nuclei also synthesize a group of proteins known as the neurophysins. These large-molecular-weight proteins appear to bind to ADH (Ginsburg and Ireland 1966), and possibly serve to neutralize the positive charge of the peptide (Sachs et al. 1969). It is believed that the peptide-neurophysin complexes are transported down the axons, in this bound form, to the posterior pituitary (Zimmerman, Carmel, Husain, Ferin, Tannenbaum, Frantz, and Robinson 1973). Whether the complexes separate before or after release, or not at all, is not known (see below).

A variety of physiological stimuli have been shown to cause ADH release, but there is still little known about the mechanisms involved in this release process. It appears that the neurosecretory cells, described above, have electrophysiological properties similar to conventional neurons. ADH release occurs when action potentials (generated in the supraoptic or paraventricular nucleus) propagate down the axons resulting in depolarization of the nerve terminals in the posterior pituitary. It is well established that hormone release involves calcium but its precise role is undecided (see reviews by Douglas 1973; Thorn, Russell, Torp-Pedersen, and Treiman 1978); membrane depolarization causes an influx of calcium and hormone release probably occurs by exocytosis. However, it is unclear whether the peptide-neurophysin complexes dissociate, thereby liberating the free hormone (Ginsburg and Ireland 1966), or if the peptide-neurophysin complexes are released intact (Douglas and Poisner 1964; Douglas 1973).

8.2. RELEASE OF ANTIDIURETIC HORMONE

8.2.1. Osmoreceptors and sodium sensors

Verney (1947) reported that injection of hypertonic saline into the carotid artery of conscious, water-loaded dogs caused a prompt and reversible decline in urine flow; this response was likened to that produced by intravenous injection of pituitary extract. Intracarotid injections of hypertonic sucrose were also effective in causing an antidiuresis, whereas hypertonic urea was not (Verney 1947). From this it was concluded that the secretion of ADH from the posterior pituitary was controlled by an intracranial osmoreceptor which responded to changes in the osmolality of blood in the cerebral vessels. Hypertonic saline and sucrose do not freely cross cell membranes whereas urea does. Thus, an increase in plasma osmolality, due to either saline or sucrose, will cause water to leave cells, whereas urea, which distributes both intra- and extracellularly, will not exert any osmotic effect. Verney (1947) suggested that the stimulus which triggered hormone release was a fall in the turgor of the osmosensitive cells, but this proposal has since been questioned. Andersson, Dallman, and Olsson (1969) confirmed that intracarotid administration of saline, sucrose, or fructose caused ADH release, but they also showed that, when these three substances were administered intraventricularly, only saline was an effective stimulus for ADH release. Indeed Eriksson (1974) found that intraventricular administration of hypertonic glucose, fructose, or sucrose inhibited ADH release. These workers suggested that there was a 'juxtaventricular' sodium sensor in the region of the third ventricle which triggered ADH release in response to an increase in sodium concentration of the cerebrospinal fluid (Andersson et al. 1969; Andersson 1978). So, when hypertonic sugar solutions are administered into the cerebral ventricles, the osmolality of the cerebospinal fluid increases but its sodium concentration decreases and ADH release is inhibited (Eriksson 1974). Such a mechanism could explain some of the results obtained by Athar and Robertson (1974) who compared the effects of intravenous infusion of hypertonic saline, mannitol, urea, or glucose on ADH release (Fig. 8.1). In their studies saline and mannitol were equally effective in causing ADH release, urea was without effect, and glucose inhibited ADH release. Saline and mannitol cause a shift of water out of the cerebrospinal fluid, thereby increasing its sodium concentration, which should stimulate ADH release (Andersson 1978). Since glucose has been shown to cross the blood-brain barrier by a selective transport process (Crone 1965a,b), it could inhibit ADH release by reducing the sodium concentration in the cerebrospinal fluid. However, the failure of urea to elicit ADH release is difficult to explain since it causes large shifts of water out of the cerebrospinal fluid (Reed and Woodbury 1962) and should cause ADH release in the same way as saline or mannitol. At present neither osmoreceptor nor sodium sensor hypotheses can account for all the experimental findings, and it may be that both mechanisms are involved in modulating ADH output.

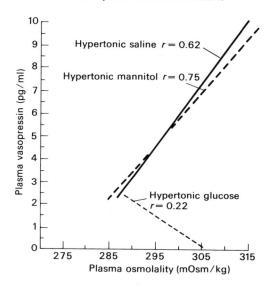

Fig. 8.1. Schematic representation of the relationship between plasma vasopressin and plasma osmolality in healthy adults during the infusion of hypertonic saline, mannitol, or glucose. In each case, the rates of change in osmolality and expansion of blood volume were the same. (With permission, Robertson et al. (1976).)

In an attempt to locate the osmoreceptive cells, Jewell and Verney (1957) studied the effects of ligating different branches of the internal carotid artery on the antidiuretic response to hypertonic saline; they concluded that the osmoreceptors were in or near the supraoptic and paraventricular nuclei (Jewell and Verney 1957), which led them to suggest that the neurosecretory cells in these hypothalamic nuclei might be the osmoreceptors. More recent studies indicate that the osmoreceptors and the neurosecretory cells might be separate elements (Hayward and Vincent 1970; Hayward and Jennings 1973; Peck and Blass 1975).

Several groups of investigators have shown that ADH release can be modified by various drugs. Bridges and Thorn (1970) found that ADH release in response to an osmotic stimulus was blocked by reserpine, phenoxybenzamine, atropine, and nicotinic receptor antagonists. While it is likely that some of the effects they observed were non-specific, Bridges and Thorn (1970) suggested that their findings could be most easily explained if the osmosensitive cell was a monoaminergic neuron synapsing with a cholinergic interneuron which stimulated ADH release from the neurosecretory cell. However, Barker, Crayton, and Nicoll (1971) demonstrated that microiontophoretic application of noradrenalin depressed the activity of supraoptic neurons, whereas application of acetylcholine was sometimes excitatory (16 per cent), and sometimes inhibitory (84 per cent). They further investigated the cholinergic mechanisms with the use of muscarinic and nicotinic agonists and antagonists and concluded that nicotinic receptor stimulation caused the excitatory effects while stimulation of muscarinic receptors was responsible for the inhibitory effects (Barker *et al.* 1971). Bhargava, Kulshrestha, and Srivastava (1972) studied the effects of intraventricular administration of adrenoceptor and cholinoceptor agonists and antagonists on plasma ADH concentration; they concluded that the effects of α-adrenoceptor and cholinoceptor stimulation were excitatory while β-adrenoceptor activation caused inhibition of ADH release. Thus the problem of the neural control of ADH release is unresolved, and the possibility remains that the neurosecretory cells are osmosensitive and their activity is modulated by various neuronal inputs.

In the earliest studies concerned with the hormonal control of water balance, ADH release was estimated indirectly by measuring urine concentration and volume (Verney 1947). Later, bioassay techniques were developed for measuring plasma levels of ADH, but none of these were very satisfactory. The development of an immunoassay (Robertson, Klein, Roth, and Gorden 1970), and later, a radioimmunoassay (Robertson, Mahr, Athar, and Sinha 1973) for ADH has led to a much broader understanding of the mechanisms controlling ADH release. Robertson *et al.* (1973) demonstrated a direct relationship between plasma osmolality and circulating ADH levels and they obtained two useful parameters from the regression line relating these variables:

(1) The point at which an increase in plasma osmolality begins to cause an increase in plasma ADH (set-point). This value is extremely consistent within a population (281 ± 0.9 mOsm/kg, mean ± SEM, $N = 16$ in normal man; Robertson, Athar, and Shelton, 1977).

(2) The slope of the line, which gives a measure of the sensitivity of the response. In normal man the sensitivity is such that a fall in osmolality of 2.9 mOsm/kg (achieved by a 1 per cent increase in body water) causes a 50 per cent reduction in the plasma ADH level

and in the urine osmolality, with the result that urine output doubles (Robertson, Shelton, and Athar 1976).

Other factors which are known to influence ADH release may do so directly and/or by changing the relationship between plasma osmolality and ADH; two such factors are considered below.

8.2.2. Volume receptors

Rydin and Verney (1938) reported that the removal of small volumes of blood from a dog inhibited water diuresis. Later Rocha e Silva and Rosenberg (1969) demonstrated that a 30-mm Hg reduction in diastolic pressure in dogs caused a fourfold increase in ADH concentration in the blood, and more recently Dunn, Brennan, Nelson, and Robertson (1973) have shown that a blood-volume depletion in excess of eight per cent stimulates ADH release in rats. A reduction in the central blood volume of man, as a result of changing from the recumbent to the upright posture, increases the circulating levels of ADH (Robertson and Athar 1976). While the evidence for volume receptors modulating ADH output is strong, there is no general agreement on the location of these receptors (Goetz, Bond, and Bloxham 1975). Left atrial distension causes a diuresis from intact and surgically denervated kidneys (Kappagoda, Linden, Snow, and Whitaker 1975) but some workers have shown a reduction in the circulating levels of ADH in response to left atrial distension (Johnson, Moore, and Segar 1969; Brennan, Malvin, Jochim, and Roberts 1971) whilst others (Kappagoda, Linden, Snow, and Whitaker 1974) have failed to show any consistent change in plasma levels of ADH. Furthermore, Kappagoda *et al.* (1975) found that the diuresis still occurred after destruction of the posterior pituitary and therefore concluded that the humoral factor was not ADH. This view is contested, however, since Torrente, Robertson, McDonald, and Schrier (1975) clearly showed that the diuresis caused by left atrial distension was directly related to a fall in the concentration of circulating ADH measured by radioimmunoassay. Furthermore, Koizumi and Yamashita (1978) recently demonstrated that left atrial stretch reduced the firing rate of neurons in the supraoptic and paraventricular nuclei (Fig. 8.2).

It has been suggested that arterial baroreceptor activity influences ADH release. Share (1967) demonstrated that ADH release during haemorrhage was prevented by inactivation of the carotid sinus

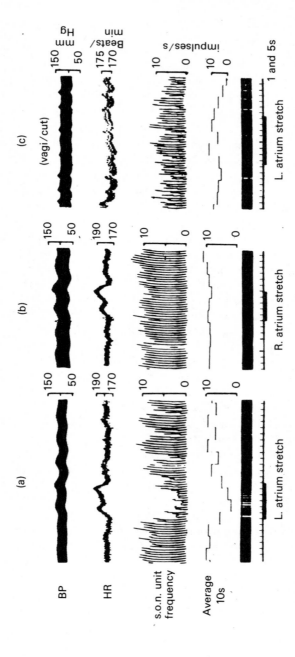

Fig. 8.2. Effects of left (a) or right (b) atrial stretch on activity in a supraoptic (s.o.n.) neuron of the dog. Right atrial stretch was without effect, while the inhibitory effect of left atrial stretch was abolished by vagal section (c). From top to bottom; systemic blood pressure (BP); heart rate (HR); integrated record of firing frequencies of a s.o.n. neuron, counted every second; average firing frequencies every 10 s; directly recorded spikes; and time. (With permission, Koizumi and Yamashita (1978).)

baroreceptors (see Chapter 2). Intravenous administration of pressor doses of noradrenalin cause a diuresis (Schrier and Berl 1973) whereas depressor doses of isoprenaline cause an antidiuresis (Levi, Grinblat, and Kleeman 1971; Schrier, Lieberman, and Ufferman 1972). These effects could not be mimicked by intrarenal administration of the drugs (Schrier, Berl, Harbottle, and McDonald 1975), were obliterated by hypophysectomy (Schrier *et al.* 1972), were absent in patients with diabetes insipidus (Berl, Harbottle, and Schrier 1974), and were abolished by baroreceptor denervation (Berl, Cadnapaphornchai, Harbottle, and Schrier 1974a,b). It was suggested, therefore, that a change in afferent vagal nerve discharge, due to altered baroreceptor activity, was responsible for the diuresis and antidiuresis through an influence on endogenous ADH release (Schrier *et al.* 1975). This hypothesis is supported by the finding that arterial baroreceptor stimulation is accompanied by a reduction in the discharge of phasically active neurons in the supraoptic nucleus (Harris 1979).

In addition to directly affecting ADH release, a change in blood-pressure or -volume can alter the relationship between plasma osmolality and ADH release. Dunn *et al.* (1973) showed that a 10-15 per cent reduction in blood volume in rats increased circulating ADH levels and also produced a parallel leftward shift of the regression line relating plasma osmolality and ADH (Fig. 8.3), such that ADH release occurred at a lower plasma osmolality; a similar situation occurs in man (Robertson and Athar 1976). Furthermore, in certain hypertensive states the slope of the line relating plasma osmolality to ADH may be reduced, and conversely in hypotension the slope may be increased (Robertson *et al.* 1976). However, rather than representing an actual change in sensitivity of the system, it is possible that alterations in the volume of distribution of ADH may be responsible for this phenomenon.

In certain clinical conditions, a patient may have a selective loss of the osmoreceptor-mediated control of ADH release with little or no impairment of the volume or pressure sensitive mechanism (Robertson *et al.* 1977). It is thus possible that there are two populations of ADH-secreting neurons, one influenced by osmotic stimuli and one by baroreceptor activity, or it may be that one population of neurons is influenced by both types of stimuli (Fig. 8.4).

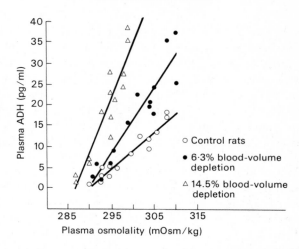

Fig. 8.3. The effect of hypovolaemia (induced by intraperitoneal injection of polyethylene glycol) on the response of plasma ADH to osmotic challenges. (With permission, Dunn *et al.* (1973).)

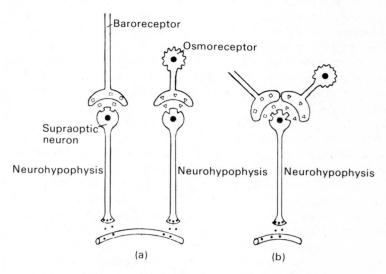

Fig. 8.4. Diagrammatic representation of the influence of baroreceptor and osmoreceptor influences on supraoptic neurons secreting ADH. There may be two separate populations which receive afferent input from baroreceptors or osmoreceptors (a), or one population influenced by both afferent inputs (b). (With permission, Robertson *et al.* (1977).)

8.2.3. Renin-angiotensin system

The central and renal renin-angiotensin systems have been mentioned in earlier chapters in the context of the control of sodium balance. Angiotensin II is also important in regulating water balance, both through its effects on drinking (see Chapter 10), and through its effects on ADH release. Bonjour and Malvin (1970) showed that intravenous infusion of a low dose of angiotensin II in conscious dogs caused a significant increase in circulating ADH levels, whilst intracarotid administration caused an even greater increase (Mouw, Bonjour, Malvin, and Vander 1971). However, Claybaugh and his associates were unable to repeat these findings. In their experiments there was no change in the plasma level of ADH when angiotensin II was infused into anaesthetized dogs (Claybaugh, Share, and Shimizu 1972) or when renin was infused into conscious dogs (Claybaugh 1976). Furthermore, the finding that the ADH response to haemorrhage was not affected by nephrectomy (Claybaugh and Share 1972) suggested to these workers that the endogenous release of renal renin was not important in ADH release. Andersson and Eriksson (1971) also found intravenous angiotensin II administration ineffective in causing ADH release in conscious goats, but they showed that intraventricular administration of angiotensin II given in conjunction with isotonic saline, caused ADH release; when hypertonic saline was used the effect was greater (Andersson and Eriksson 1971). In contrast, there was no ADH release when angiotensin II was administered with isotonic glucose (Andersson 1977). It was suggested that angiotensin II might act centrally in some way to facilitate the excitation of the juxtaventricular sodium sensor (Andersson 1977, 1978) and thereby cause ADH release. The synergistic action of central angiotensin II administration and hypertonic saline was also demonstrated in dogs by Shimizu, Share, and Claybaugh (1973). These workers measured the release of ADH in response to the osmotic challenge of hypertonic saline in the presence and in the absence of angiotensin II (Fig. 8.5). The effect of angiotensin II was to increase the slope of the regression line relating plasma osmolality to ADH, i.e. to increase the sensitivity of the response (Shimizu *et al.* 1973). The observation that simultaneous perfusion of the cerebral ventricles with angiotensin II and indomethacin (a prostaglandin synthetase inhibitor) substantially reduced the effect of angiotensin II on ADH release (Yamamoto, Share, and Shade 1976) has led to the speculation that angiotensin II exerts its effect on ADH release by way

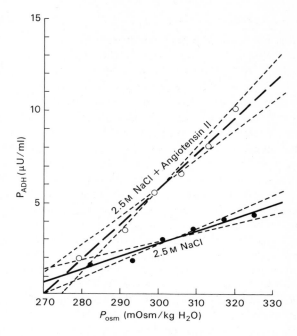

Fig. 8.5. Effect of i.v. infusion of 2.5M NaCl alone or with intracarotid infusion of angiotensin II on plasma ADH concentration (P_{ADH}). Each point represents the mean of the plasma ADH concentrations in each group of experiments plotted against the mean of their plasma osmolalities. The average linear regressions were drawn for the control group (solid line) and angiotensin group (dotted heavy line). 95 per cent confidence limits on the two average slopes are shown by the dotted lighter lines. (With permission, Shimizu *et al.* (1973).)

of a central action mediated by prostaglandins (Share, Claybaugh, Shimizu, Yamamoto, and Shade 1978).

8.3. ACTIONS OF ANTIDIURETIC HORMONE

The principal action of ADH in the control of cardiovascular homeostasis is its ability to alter the water and urea permeability of the renal collecting duct and thereby modulate urine concentration and volume (see Chapter 9). The mechanism by which ADH alters the permeability properties of epithelia is not completely understood, although it is now generally agreed that the initial intracellular events involve binding of the hormone to the receptor sites on the basal membrane, activation of adenyl cyclase, and intracellular generation of cyclic AMP (cAMP; see review Hays 1976). The intracellular cAMP is then destroyed by the action of phosphodiesterase and inactive 5' AMP is formed. The magnitude of the permeability change depends directly on the intracellular level of cAMP. Several humoral factors affect the intracellular production of cAMP and hence influence the action of ADH.

The luminal surface of the collecting duct is, indisputably, the final site of action of the ADH-mediated effect on membrane permeability (Grantham and Burg 1966; Shafer and Andreoli 1972). However, the mechanism whereby intracellular cAMP causes a change in luminal membrane permeability is unclear. One possibility is that it activates a kinase which phosphorylates a membrane protein (Schwartz, Shlatz, Kinne-Saffran, and Kinne 1974), although some investigators have shown that ADH or cAMP decrease protein phosphorylation (Handler, Strewler, and Orloff 1977). Thus the direction of change in phosphoprotein metabolism and its role in the action of ADH is undecided. Alternatively, luminal membrane permeability may be modified through a change in microtubule or microfilament formation. Dousa and Barnes (1974) studied the effects of antimitotic agents (which inhibit microtubule assembly) on water diuresis in conscious rats and showed that, although these agents were ineffective when administered alone, they inhibited the effect of exogenously administered ADH. Similar findings were reported by Taylor, Maffly, Wilson, and Reaven (1975). Levine, Franki, and Hays (1973) found that phloretin inhibited changes in collecting duct permeability to urea irrespective of whether these were caused by exogenous ADH or cAMP; this indicates that the site of action of phloretin is distal to the formation of cAMP. Since other agents (e.g. methylene blue) only affected the ADH-mediated changes in urea permeability their action must have been proximal to cAMP formation. Neither phloretin nor methylene blue influenced water movements. Levine, Levine, Worthington, and Hays (1976) observed that antimitotic agents (colchicine or vinblastine) blocked the effects of exogenous ADH or cAMP on collecting duct permeability to water, while halothane interfered only with the effects of ADH. Since halothane does not affect urea movements, and methylene blue does not affect water movements, it has been suggested that two separate cAMP systems might exist — one mediating the effects of ADH on water permeability (possibly by an influence on microtubules) and the other concerned with the effect of ADH on urea permeability (for review of evidence see Hays 1976).

Apart from the renal actions described above there is evidence that the pressor action of ADH may be of physiological importance in controlling arterial blood-pressure. Pressor quantities of ADH are released in response to haemorrhage (Rocha e Silva and Rosenberg 1969) and in response to intraventricular administration of angiotensin II or carbachol (Hoffman, Phillips, Schmid, Falcon, and Weet 1977). Furthermore, physiological doses of ADH enhance the pressor responses to catecholamines in the pithed rat preparation (Bartelstone and Nasmyth 1965). Thus, it is possible that endogenously liberated ADH may act to maintain arterial blood-pressure. Möhring and his colleagues have suggested that this pressor action of ADH may be important in the development of hypertension in rats given deoxycorticosterone acetate and saline (Möhring, Möhring, Petri, and Haack 1977) and in spontaneously hypertensive rats (Möhring, Kintz, and Schoun 1978).

REFERENCES

Andersson, B. (1977). Regulation of body fluids. *Ann. Rev. Physiol.* **39**, 185–200.

—— (1978). Sodium versus osmotic sensitivity in cerebral control of water balance. In *Osmotic and volume regulation. Proc. Alfred Benzon Symposium XI* (ed. C. B. Jorgensen and E. Skadhauge), pp. 84–94. Academic Press, London.

——, Dallman, M. F., and Olsson, K. (1969). Observations on central control of drinking and of the release of antidiuretic hormone (ADH). *Life Sci.* **8**, 425–32.

—— and Eriksson, L. (1971). Conjoint action of sodium and angiotensin on brain mechanisms controlling water and salt balances. *Acta physiol. Scand.* **81**, 18–29.

Athar, S. and Robertson, G. L. (1974). Osmotic control of vasopressin secretion in man. *Clin. Res.* **22**, 335A.

Barker, J. L., Crayton, J. W., and Nicoll, R. A. (1971). Supraoptic neurosecretory cells: adrenergic and cholinergic sensitivity. *Science* **171**, 208–10.

Bartelstone, H. J. and Nasmyth, P. A. (1965). Vasopressin potentiation of catecholamine actions in dog, rat, cat and rat aortic strip. *Amer. J. Physiol.* **208**, 754–62.

Berl, T., Cadnapaphornchai, P., Harbottle, J. A., and Schrier, R. W. (1974a). Mechanism of suppression of vasopressin during alpha-adrenergic stimulation with norepinephrine. *J. clin. Invest.* **53**, 219–27.

——, ——, ——, —— (1974b). Mechanism of stimulation of vasopressin release during beta-adrenergic stimulation with isoproterenol. *J. clin. Invest.* **53**, 857–67.

―, Harbottle, J. A., and Schrier, R. W. (1974). Effect of alpha- and beta-adrenergic stimulation on renal water excretion in normal subjects and patients with diabetes insipidus. *Kidney Int.* **6**, 247–53.

Bhargava, K. P., Kulshrestha, V. K., and Srivastava, Y. P. (1972). Central cholinergic and adrenergic mechanisms in the release of antidiuretic hormone. *Brit. J. Pharmacol.* **44**, 617–27.

Bonjour, J. P. and Malvin, R. L. (1970). Stimulation of ADH release by the renin–angiotensin system. *Amer. J. Physiol.* **218**, 1555–9.

Brennan, L. A., Jr., Malvin, R. L., Jochim, K. E., and Roberts, D. E. (1971). Influence of right and left atrial receptors on plasma concentrations of ADH and renin. *Amer. J. Physiol.* **221**, 273–8.

Bridges, T. and Thorn, N. A. (1970). The effect of autonomic blocking agents on vasopressin release *in vivo* induced by osmoreceptor stimulation. *J. Endocrinol.* **48**, 265–76.

Claybaugh, J. R. (1976). Effect of dehydration on stimulation of ADH release by heterologous renin infusions in conscious dogs. *Amer. J. Physiol.* **231**, 655–60.

―and Share, L. (1972). Role of the renin–angiotensin system in the vasopressin response to hemorrhage. *Endocrinol.* **90**, 453–60.

―, ―, and Shimizu, K. (1972). The inability of infusions of angiotensin to elevate the plasma vasopressin concentration in the anaesthetized dog. *Endocrinol.* **90**, 1647–52.

Crone, C. (1965a). Facilitated transfer of glucose from blood into brain tissue. *J. Physiol.* **181**, 103–13.

―(1965b). Permeability of brain capillaries to non-electrolytes. *Acta physiol. Scand.* **64**, 407–17.

Douglas, W. W. (1973). How do neurones secrete peptides? Exocytosis and its consequences, including 'synaptic vesicle' formation, in the hypothalamo-neurohypophyseal system. *Prog. Brain Res.* **39**, 21–39.

―and Poisner, A. N. (1964). Stimulus-secretion coupling in a neurosecretory organ: the role of calcium in the release of vasopressin from the neurohypophysis. *J. Physiol.* **172**, 1–18.

Dousa, T. P. and Barnes, L. D. (1974). Effects of colchicine and vinblastine on the cellular action of vasopressin in mammalian kidney. A possible role of microtubules. *J. clin. Invest.* **54**, 252–62.

Dunn, F. L., Brennan, T. J., Nelson, A. E., and Robertson, G. L. (1973). The role of blood osmolality and volume in regulating vasopressin secretion in the rat. *J. clin. Invest.* **52**, 3212–19.

Eriksson, L. (1974). Effect of lowered CSF sodium concentration on the central control of fluid balance. *Acta physiol. Scand.* **91**, 61–8.

Gauer, O. H. (1968). Osmocontrol versus volume control. *Fed. Proc.* **27**, 1132–6.

Ginsburg, M. and Ireland, M. (1966). The role of neurophysin in the transport and release of neurohypophyseal hormones. *J. Endocrinol.* **35**, 289–98.

Goetz, K. L., Bond, G. C., and Bloxham, D. D. (1975). Atrial receptors and renal function. *Physiol. Rev.* **55**, 157-205.

Grantham, J. J. and Burg, M. B. (1966). Effect of vasopressin and cyclic AMP on permeability of isolated collecting tubules. *Amer. J. Physiol.* **211**, 255-9.

Handler, J. S., Strewler, G. J., and Orloff, J. (1977). Role of protein synthesis and phosphoprotein metabolism in cellular response to vasopressin. In *Disturbances in body fluid osmolality* (ed. T. E. Andreoli, J. J. Grantham, and F. C. Rector, Jr.), pp. 85-95. American Physiological Society, Bethesda, Maryland.

Harris, M. C. (1979). Effects of chemoreceptor and baroreceptor stimulation on the discharge of hypothalamic supraoptic neurones in rats. *J. Endocrinol.* **82**, 115-25.

Hays, R. M. (1976). Antidiuretic hormone. *New Engl. J. Med.* **295**, 659-65.

Hayward, J. N. and Jennings, D. P. (1973). Activity of magnocellular neuroendocrine cells in the hypothalamus of unanesthetized monkeys. II. Osmosensitivity of functional cell types in the supraoptic nucleus and the internuclear zone. *J. Physiol.* **232**, 545-72.

―― and Vincent, J. D. (1970). Osmosensitive single neurones in the hypothalamus of unanaesthetized monkeys. *J. Physiol.* **210**, 947-72.

Hoffman, W. E., Phillips, M. I., Schmid, P. G., Falcon, J., and Weet, J. F. (1977). Antidiuretic hormone release and the pressor response to central angiotensin II and cholinergic stimulation. *Neuropharmacol.* **16**, 463-72.

Jewell, P. A. and Verney, E. B. (1957). An experimental attempt to determine the site of neurohypophyseal osmoreceptors in the dog. *Phil. Trans. roy. Soc. Ser. B* **240**, 197-324.

Johnson, J. A., Moore, W. W., and Segar, W. E. (1969). Small changes in left atrial pressure and plasma antidiuretic hormone titers in dogs. *Amer. J. Physiol.* **217**, 210-14.

Kappagoda, C. T., Linden, R. J., Snow, H. M., and Whitaker, E. M. (1974). Left atrial receptors and the antidiuretic hormone. *J. Physiol.* **237**, 663-83.

――, ――, ――, and ―― (1975). Effect of destruction of the posterior pituitary on the diuresis from left atrial receptors. *J. Physiol.* **244**, 757-70.

Koizumi, K. and Yamashita, H. (1978). Influence of atrial stretch receptors on hypothalamic neurosecretory neurones. *J. Physiol.* **285**, 341-58.

Levi, J., Grinblat, J., and Kleeman, C. R. (1971). Effect of isoproterenol on water diuresis in rats with congenital diabetes insipidus. *Amer. J. Physiol.* **221**, 1728-32.

Levine, S. D., Franki, N., and Hays, R. M. (1973). Effect of phloretin on water and solute movement in the toad bladder. *J. clin. Invest.* **52**, 1435-42.

―――, Levine, R. D., Worthington, R. E., and Hays, R. M. (1976). Selective inhibition of osmotic water flow by general anaesthetics in toad urinary bladder. *J. clin. Invest.* **58**, 980–8.

Möhring, J., Kintz, J., and Schoun, J. (1978). Role of vasopressin in blood pressure control of spontaneously hypertensive rats. *Clin. Sci. Molec. Med.* **55** (Suppl. 4), 247S–250S.

―――, Möhring, B., Petri, M., and Haack, D. (1977). Vasopressor role of ADH in the pathogenesis of malignant DOC hypertension. *Amer. J. Physiol.* **232**, F260–F269.

Mouw, D., Bonjour, J. P., Malvin, R. L., and Vander, A. (1971). Central action of angiotensin in stimulating ADH release. *Amer. J. Physiol.* **220**, 239–42.

Peck, J. W. and Blass, E. M. (1975). Localization of thirst and antidiuretic osmoreceptors by intracranial injections in rats. *Amer. J. Physiol.* **228**, 1501–9.

Reed, D. J. and Woodbury, D. M. (1962). Effect of hypertonic urea on cerebrospinal fluid pressure and brain volume. *J. Physiol.* **164**, 252–64.

Robertson, G. L. and Athar, S. (1976). The interaction of blood osmolality and blood volume in regulating plasma vasopressin in man. *J. clin. Endocrinol. Metab.* **42**, 613–20.

―――, ―――, and Shelton, R. L. (1977). Osmotic control of vasopressin function. In *Disturbances in body fluid osmolality* (ed. T. E. Andreoli, J. J. Grantham, and F. C. Rector, Jr.), pp. 125–48. American Physiological Society, Bethesda, Maryland.

―――, Klein, L. A., Roth, J., and Gorden, P. (1970). Immunoassay of plasma vasopressin in man. *Proc. Nat. Acad. Sci. USA* **66**, 1298–305.

―――, Mahr, E. A., Athar, S., and Sinha, T. (1973). Development and clinical application of a new method for the radioimmunoassay of arginine vasopressin in human plasma. *J. clin. Invest.* **52**, 2340–52.

―――, Shelton, R. L., and Athar, S. (1976). The osmoregulation of vasopressin. *Kidney Int.* **10**, 25–37.

Rocha e Silva, M. Jr. and Rosenberg, M. (1969). The release of vasopressin in response to haemorrhage and its role in the mechanism of blood pressure regulation. *J. Physiol.* **202**, 535–57.

Rydin, H. and Verney, E. B. (1938). The inhibition of water diuresis by emotional stress and muscular exercise. *Quart. J. exp. Physiol.* **27**, 343–74.

Sachs, H. (1967). Biosynthesis and release of vasopressin. *Amer. J. Med.* **42**, 687–700.

―――(1969). Neurosecretion. *Advan. Enzymol.* **32**, 327–72.

―――, Fawcett, P., Takabatake, Y., and Portanova, R. (1969). Biosynthesis and release of vasopressin and neurophysin. *Recent Prog. Hormone Res.* **25**, 447–84.

Schafer, J. A. and Andreoli, T. E. (1972). Cellular constraints to diffusion. The effect of antidiuretic hormone on water flows in isolated mammalian collecting tubules. *J. clin. Invest.* **51**, 1264–78.

Schrier, R. W. and Berl, T. (1973). Mechanism of effect of α-adrenergic stimulation with norepinephrine on renal water excretion. *J. clin. Invest.* **52**, 502–11.

——, ——, Harbottle, J. A., and McDonald, K. M. (1975) Catecholamines and renal water excretion. *Nephron* **15**, 186–96.

Schrier, R. W., Lieberman, R., and Ufferman, R. C. (1972). Mechanism of antidiuretic effect of beta adrenergic stimulation. *J. clin. Invest.* **51**, 97–111.

Schwartz, I. L., Shlatz, L. J., Kinne-Saffran, E., and Kinne, R. (1974). Target cell polarity and membrane phosphorylation in relation to the mechanism of action of antidiuretic hormone. *Proc. Nat. Acad. Sci* **71**, 2595–9.

Share, L. (1967). Role of peripheral receptors in the increased release of vasopressin in response to hemorrhage. *Endocrinol.* **81**, 1140–6.

——, Claybaugh, J. R., Shimizu, K., Yamamoto, M., and Shade, R. E. (1978). Role of the renin–angiotensin system and the prostaglandins in the control of vasopressin release. In *Osmotic and volume regulation. Alfred Benzon Symposium XI* (ed. C. B. Jorgensen and E. Skadhauge), pp. 248–57. Academic Press, New York.

Shimizu, K., Share, L., and Claybaugh, J. R. (1973). Potentiation by angiotensin II of the vasopressin response to an increasing plasma osmolality. *Endocrinol.* **93**, 42–50.

Takabatake, Y. and Sachs, H. (1964). Vasopressin biosynthesis. III. In vitro studies. *Endocrinol.* **75**, 934–42.

Taylor, A., Maffly, R., Wilson, L., and Reaven, E. (1975). Evidence for involvement of microtubules in the action of vasopressin. *Ann. NY Acad. Sci.* **253**, 723–37.

Thorn, N. A., Russell, J. T., Torp-Pedersen, C., and Treiman, M. (1978). Calcium and neurosecretion. *Ann. NY Acad. Sci.* **367**, 618–38.

Torrente, A. De., Robertson, G. L., McDonald, K. M., and Schrier, R. W. (1975). Mechanism of diuretic response to increased left atrial pressure in anaesthetized dog. *Kidney Int.* **8**, 355–61.

Verney, E. B. (1947). The antidiuretic hormone and the factors which determine its release. *Proc. roy. Soc. Ser. B* **135**, 25–106.

Yamamoto, M., Share, L., and Shade, R. E. (1976). Effect of ventriculocisternal perfusion with angiotensin II and indomethacin on the plasma vasopressin concentration. *Fed. Proc.* **35**, 690 (abst. 2668).

Young, D. B., Pan, Y-J., and Guyton, A. C. (1977). Control of extracellular sodium concentration by antidiuretic hormone–thirst feedback mechanism. *Amer. J. Physiol.* **232**, R145–R149.

Zimmerman, E. A., Carmel, P. W., Husain, M. K., Ferin, M., Tannenbaum, M., Frantz, A. G., and Robinson, A. G. (1973). Vasopressin and neurophysin: high concentrations in monkey hypophysial portal blood. *Science* **182**, 925–7.

9. Urine concentration and dilution

As mentioned in the previous chapter, the principal action of antidiuretic hormone (ADH) is to increase the permeability of the collecting duct to water and urea and thereby modulate the concentration of urine. However, if these changes in permeability were the only determinants of the final concentration of the urine, then urine could never be more concentrated than plasma. Clearly some other mechanism must operate to achieve the hypertonic urine which needs to be produced under certain circumstances.

For many years it has been known that there is a progressive rise in the tissue osmolality from the corticomedullary junction to the papillary tip of the kidney (Wirz, Hargitay, and Kuhn 1951), and that the ratio of cortical to medullary thickness is directly related to the maximum achievable urine concentration (Schmidt-Nielsen and O'Dell 1961). It is now universally accepted that the process of concentrating the urine depends on osmotic equilibration across vascular and tubular elements within the renal medulla; the operation of this process depends critically on the membrane properties of the different segments of the renal tubule discussed below (see also Andreoli, Berliner, Kokko, and Marsh 1978).

9.1. TUBULAR CHARACTERISTICS

(1) The descending limb of the loop of Henle (DHL) is freely permeable to water in all species studied (for references see Jamison 1976). In the rabbit, the solute permeability is very low (Kokko 1970, 1972), whereas in both the rat (Morgan and Berliner 1968) and the desert rat, Psammomys (de Rouffignac and Morel 1969), there is a relatively high permeability to solutes in the DLH.

(2) The thin ascending limb of the loop of Henle (tALH) is impermeable to water in both the rat (Morgan and Berliner 1968) and the rabbit (Kokko and Rector 1972; Imai and Kokko 1974). Imai and Kokko (1974) found that the tALH of the rabbit was more permeable to sodium (15-fold) and urea (4-fold) than the DLH, whereas Morgan and Berliner (1978) reported that the sodium permeability of the tALH of the rat was five times less than that of the DLH. There is some evidence that sodium may be actively reabsorbed in the tALH of the

rat (Jamison, Bennett, and Berliner 1967) and the hamster (Marsh 1970; Marsh and Azen 1975; Marsh and Martin 1977), but Imai and Kokko (1974) were unable to demonstrate active transport in this segment of the rabbit tubule.

(3) The thick ascending limb of the loop of Henle (TALH) is impermeable to both water and urea (Jamison 1976; Kokko and Tisher 1976). There is an active transport process in the TALH (Burg and Green 1973; Burg and Stoner 1974; see Chapter 5) which provides the driving force for the renal countercurrent mechanism described below.

(4) The permeability of the collecting duct to water and urea is dependent upon circulating ADH levels; this property is of pivotal importance in determining the final concentration of the urine. ADH increases the water permeability of the cortical and medullary collecting duct segments (Gross, Imai, and Kokko 1975; Kokko and Tisher 1976), whereas the permeability to urea is only increased in the region of the inner medulla at a level which corresponds to the tips of the loops of Henle (Morgan and Berliner 1968).

9.2. RENAL COUNTERCURRENT MECHANISMS

In 1972, two groups, working independently, developed models to describe the renal countercurrent mechanism for the formation of concentrated urine (Kokko and Rector 1972; Stephenson 1972). The characteristics of these two models are essentially the same except that the one described by Kokko and Rector (1972) is based solely on the permeability properties of the rabbit kidney, whereas the one described by Stephenson (1972) allows some flexibility to account for the different tubular characteristics reported in other species (see above). The model of Kokko and Rector will be considered first (Fig. 9.1).

The driving force for the system resides in an active chloride transport out of the TALH. Chloride (and the associated sodium) movement out of the TALH increases the osmolality of the interstitium in this region (through which the DLH passes). Fluid leaving the proximal convoluted tubule is invariably isotonic, but urea may be secreted by the straight segment of the proximal tubule (Kawamura and Kokko 1976), in which case the fluid entering the DLH would be slightly hypertonic. However, the interstitial osmolality is greater and the high

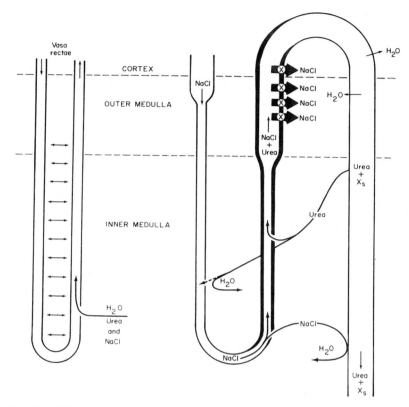

Fig. 9.1. Schematic representation of the renal countercurrent multiplication system without solute transport in the medulla. X_s = non-reabsorbable solute. For details see text. (With permission, Kokko and Rector (1972).)

water permeability and negligible solute permeability of the DLH (see above (1)) results in an osmotic withdrawal of water from the descending limb into the interstitium. Water abstraction from the DLH causes the tubular fluid to become progressively hypertonic as it approaches the tip of the loop of Henle. If there was no active transport mechanism in the tALH (Kokko and Rector 1972; Imai and Kokko 1974) then it might be argued that once the DLH reaches this level in the medulla, there would be no basis for further water removal — indeed, water should enter from the interstitium to dilute the hypertonic fluid in the tubule. However, at the level of the inner medulla the collecting duct is permeable to urea. An influx of urea at this point (see below) increases the interstitial osmolality in the region of the tip of the loop of Henle. Thus, the osmotic gradient for water abstraction is achieved by sodium and chloride in the outer medullary zones whereas at the depth of the inner medulla urea is the major osmotic component. At the tip of the loop of Henle, the interstitium and tubular fluid are in osmotic equilibrium, with sodium chloride being the major component in the tubules and urea in the interstitium.

In the rabbit, the tALH is impermeable to water, slightly permeable to urea, and highly permeable to sodium chloride (see (2) above). Hence, as fluid progresses up this portion of the nephron, sodium chloride will move out of the tubule down its concentration gradient and urea will move in, although to a lesser degree; the tonicity of the fluid reaching the TALH is therefore reduced to some extent. Active chloride transport out of the TALH means that fluid reaching the distal convoluted tubule is invariably hypotonic. It is at this stage that the involvement of ADH is important. In the absence of ADH, further solute reabsorption continues (possibly under the influence of aldosterone) but water cannot accompany the ions and urine with an osmolality far less than plasma is produced. However, in the presence of ADH there is an increase in the water permeability along the entire length of the collecting duct as it proceeds through the cortical and medullary tissue to reach the papilla. Since the osmolality of the medullary interstitium becomes progressively greater towards the region of the papilla then, under the influence of ADH, water moves out of the collecting duct down its concentration gradient; this causes the luminal concentration of urea to rise (since this region is impermeable to urea). Subsequently, when the fluid reaches the medullary level (which is permeable to urea) urea moves into the interstitium down its concentration gradient and thereby elevates the osmolality.

Urea is therefore of major importance in the renal countercurrent mechanism described by Kokko and Rector (1972). If urea were not involved, the sodium chloride gradient could not be developed in the DLH and thus the tALH would not add salt to the interstitium: under those conditions the urine concentration would only reflect the work capacity of the chloride pump in the TALH (Kokko 1977).

The work of Stephenson (1972) describes two possible models for the operation of the renal countercurrent system. The 'passive' model is identical to that which is described above. The 'active' model incorporates the possibility of active transport of sodium chloride out of the tALH. Under those conditions the role of urea changes; a high urea concentration in the inner medulla would reduce the sodium chloride concentration in that region and hence aid net reabsorption by preventing backleak (Stephenson 1972).

In summary, therefore, although the passive mechanisms described by Kokko and Rector (1972) fit well with the experimental data in the rabbit there appear to be some species differences with regard to nephron permeabilities which prevent direct extrapolation of the model to other mammals.

9.3. COUNTERCURRENT EXCHANGE

The mechanism described above demonstrates how urine more or less concentrated than plasma can be produced. However, in order for such a system to operate efficiently, the medullary concentration gradients must be maintained; the operation of this process depends on the anatomy of the renal vasculature. Post-glomerular efferent arterioles of the juxtamedullary nephrons branch to form capillaries (descending vasa recta) which closely follow the course of the DLH. At various levels in the medulla the descending vasa recta break into a capillary plexus which surrounds a portion of the ascending limb of the loop of Henle and collecting duct. The ascending vasa recta which emerge from this plexus then retrace the route of the descending capillaries (i.e. juxtaposed to the DLH) and coalesce as collecting veins before entering the arcuate veins (Kriz and Lever 1969). The capillary endothelium is freely permeable to water, slightly less permeable to solutes, and impermeable to protein (Jamison and Maffly 1976). As the descending vasa recta course through the hypertonic medullary interstitium the transcapillary hydrostatic pressure difference favours fluid loss while the colloid osmotic pressure exerted by the plasma proteins favours fluid reabsorption. Under such conditions it is difficult to conceive the direction (if any) of net fluid movement. However, due to the rapidity of blood-flow in the vasa recta there is some lag in osmotic equilibration between the interstitium and the capillaries such that the solute concentration is slightly higher in the interstitium than in the plasma at the same medullary level. This provides a transcapillary osmotic driving force which favours net fluid loss (Sanjana, Johnston, Robertson, and Jamison 1976) although solute entry into the descending vasa recta also contributes to the osmotic equilibration. If the vasa recta were to leave the medulla at this point, the hypertonic plasma would effectively 'wash out' the solutes and medullary osmolality would fall. However, the pathway followed by the ascending vasa recta is such that the blood then flows back up through the medulla to the cortex. The capillary forces — both hydrostatic and colloid osmotic — favour fluid reabsorption into the ascending vasa recta. Concurrently, solutes leave the ascending vasa recta plasma and enter the interstitium. Thus the capillaries act as 'countercurrent exchangers', providing a means by which solutes are effectively 'trapped' in the medullary interstitium and the reabsorbed water is returned to the circulation.

It is possible that ADH affects the final concentration of the urine, not only by its effects on water and urea permeability of the collecting ducts (see Chapter 8) but also by influencing the distribution of renal blood-flow. Physiological increments in circulating ADH levels increase outer medullary (Fischer, Grünfeld, and Barger 1970) and inner cortical (Johnson, Park, and Malvin 1977) blood-flow and concomitantly decrease outer cortical flow (Johnson *et al.* 1977). It has been suggested that these effects may be due either to an effect of ADH on the renal vasculature or to the release of some vasoactive substances, possibly prostaglandins (see Johnson *et al.* 1977). Such changes would re-direct the blood away from the superficial nephrons and towards the juxtamedullary nephrons. Since the latter have longer loops of Henle then it might be assumed that they would be more effective in concentrating the urine; this conclusion is, however, debatable (Martinez-Maldonado and Opava-Stitzer 1978).

Davis and Schnermann (1971) described an alternative mechanism by which ADH could alter renal blood-flow and medullary concentrating ability. In the presence of ADH, solutes and water are largely reabsorbed from the distal tubule and proximal collecting duct into the cortex. As a result, the flow rate in the more distal portions of the collecting duct is relatively low and thus a smaller volume of fluid enters the medulla during antidiuresis than in maximal water diuresis (this paradoxical situation has been confirmed; Jamison, Buerkert, and Lacy 1971). With reduced water entry, the medullary osmolality increases and causes the erythrocytes in the surrounding vasa recta to crenate; thus the apparent viscosity of the blood increases and postglomerular resistance rises. The effect of this is twofold. Firstly, the effective filtration pressure increases, thereby increasing juxtamedullary glomerular filtration rate and hence solute delivery to the ascending limbs of Henle. Under those conditions the efficiency of the countercurrent system increases since the rate of active solute transport in the TALH (which drives the process) is load-dependent (see p. 80). Secondly, an increased post-glomerular resistance reduces vasa recta blood-flow, reduces the washout of the medullary solutes from the interstitium, and thereby increases the efficiency of the countercurrent exchanger.

REFERENCES

Andreoli, T. E., Berliner, R. W., Kokko, J. P., and Marsh, D. J. (1978). Questions and replies: renal mechanisms for urinary concentrating and diluting processes. *Amer. J. Physiol.* **235**, F1–F11.

Burg, M. B. and Green, N. (1973). Functions of the thick ascending limb of Henle's loop. *Amer. J. Physiol.* **224**, 659–68.

—— and Stoner, L. (1974). Sodium transport in the distal nephron. *Fed. Proc.* **33**, 31–6.

Davis, J. M. and Schnermann, J. (1971). The effect of antidiuretic hormone on the distribution of nephron filtration rates in rats with hereditary diabetes insipidus. *Pflügers' Arch.* **330**, 323–34.

Fischer, R. D., Grünfeld, J. P., and Barger, A. C. (1970). Intrarenal distribution of blood flow in diabetes insipidus: role of ADH. *Amer. J. Physiol.* **219**, 1348–58.

Gross, J. B., Imai, M., and Kokko, J. P. (1975). A functional comparison of the cortical collecting tubule and the distal convoluted tubule. *J. clin. Invest.* **55**, 1284–94.

Imai, M. and Kokko, J. P. (1974). Sodium chloride, urea and water transport in the thin ascending limb of Henle. Generation of osmotic gradients by passive diffusion of solutes. *J. clin. Invest.* **53**, 393–402.

Jamison, R. L. (1976). Urinary concentration and dilution. The role of antidiuretic hormone and the role of urea. In *The kidney*, Vol. 1 (ed. B. M. Brenner and F. C. Rector), pp. 391–441. W. B. Saunders, Philadelphia, London, and Toronto.

——, Bennett, C. M., and Berliner, R. W. (1967). Countercurrent multiplication by the thin loops of Henle. *Amer. J. Physiol.* **212**, 357–66.

——, Buerkert, J., and Lacy, F. (1971). A micropuncture study of collecting tubule function in rats with hereditary diabetes insipidus. *J. clin. Invest.* **50**, 2444–52.

—— and Maffly, R. H. (1976). The urinary concentrating mechanism. *New Engl. J. Med.* **295**, 1059–67.

Johnson, M. D., Park, C. S., and Malvin, R. L. (1977). Antidiuretic hormone and the distribution of renal cortical blood flow. *Amer. J. Physiol.* **232**, F111–F116.

Kawamura, S. and Kokko, J. P. (1976). Urea secretion by the straight segment of the proximal tubule. *J. clin. Invest.* **58**, 604–12.

Kokko, J. P. (1970). Sodium chloride and water transport in the descending limb of Henle. *J. clin. Invest.* **49**, 1838–46.

—— (1972). Urea transport in the proximal tubule and the descending limb of Henle. *J. clin. Invest.* **51**, 1999–2008.

—— (1977). The role of the renal concentrating mechanisms in the regulation of serum sodium concentration. *Amer. J. Med.* **62**, 165–9.

—— and Rector, F. C. (1972). Countercurrent multiplication system without active transport in inner medulla. *Kidney Int.* **2**, 214–23.

—— and Tisher, C. C. (1976). Water movement across nephron segments involved with the countercurrent multiplication system. *Kidney Int.* **10**, 64–81.

Kriz, W. and Lever, A. F. (1969). Renal countercurrent mechanisms: structure and function. *Amer. Heart J.* **78**, 101–18.

Marsh, D. J. (1970). Solute and water flows in thin limbs of Henle's loop in the hamster kidney. *Amer. J. Physiol.* **218**, 824–31.

—— and Azen, S. P. (1975). Mechanism of NaCl reabsorption by hamster thin ascending limbs of Henle's loop. *Amer. J. Physiol.* **228**, 71–9.

—— and Martin, C. M. (1977). Origin of electrical PD's in hamster thin ascending limbs of Henle's loop. *Amer. J. Physiol.* **232**, F348–F357.

Martinez-Maldonado, M. and Opava-Stitzer, S. (1978). Urine concentration and dilution in the rat: Contribution of papillary structures during high rates of urine flow. *Kidney Int.* **13**, 194–200.

Morgan, T. and Berliner, R. W. (1968). Permeability of the loop of Henle, vasa recta, and collecting duct to water, urea and sodium. *Amer. J. Physiol.* **215**, 108–15.

Rouffignac, C. De. and Morel, F. (1969). Micropuncture study of water, electrolytes and urea movements along the loops of Henle in Psammomys. *J. clin. Invest.* **48**, 474–86.

Sanjana, V. M., Johnston, P. A., Robertson, C. R., and Jamison, R. L. (1976). An examination of transcapillary water flux in renal inner medulla. *Amer. J. Physiol.* **231**, 313–18.

Schmidt-Nielsen, B. and O'Dell, R. (1961). Structure and concentrating mechanism in the mammalian kidney. *Amer. J. Physiol.* **200**, 1119–24.

Stephenson, J. L. (1972). Concentration of urine in a central core model of the renal counterflow system. *Kidney Int.* **2**, 85–94.

Wirz, H., Hargitay, B., and Kuhn, W. (1951). Lokalisation des Konzentrierungsprozesses in der Niere durch direkte Kryoskopie. *Helv. physiol. pharmacol. Acta* **9**, 196–207.

10. Thirst

The present chapter deals with the control of drinking since, in the event of fluid loss, extracellular fluid volume is most effectively restored by increasing intake.

10.1. CELLULAR DEHYDRATION

In 1937 Gilman showed that infusion of a hypertonic saline solution into dogs caused them to drink more than did an infusion of an equiosmolar solution of urea (Gilman 1937). Since saline does not readily penetrate cells, whereas urea does (see Chapter 8), it was suggested that cellular dehydration was responsible for stimulating thirst (Gilman 1937). A reduction in blood volume (Szczepańska-Sadowska, Kozlowski, and Sobacińska 1974; Kozlowski and Szczepańska-Sadowska 1975) and an increase in circulating levels of angiotensin II (Kozlowski, Drzewiecki, and Zurawski 1972) both lower the thirst threshold to an osmotic stimulus, such that drinking occurs at a lower value of plasma osmolality (c.f. release of ADH, Chapter 8). The effect may be mediated by ADH, since volume depletion and angiotensin are both effective stimuli for ADH release (Chapter 8), and intravenous infusion of ADH can itself lower the thirst threshold (Kozlowski and Szczepańska-Sadowska 1975). Moreover, cervical vagotomy lowers the threshold of an osmotic stimulus (Kozlowski and Szczepańska-Sadowska 1975) and also causes a marked increase in plasma ADH levels (Thames and Schmid 1979). For this reason, it has been suggested that cardiothoracic receptors, possibly in the left atrium, are involved in the control of drinking.

Andersson and his colleagues (Andersson, Dallman, and Olsson 1969; Olsson 1972; Andersson 1978) proposed that there was a juxtaventricular sodium sensor responsible for thirst as well as ADH release (see Chapter 8). This hypothesis, however, is still contested and the majority of evidence is in favour of osmosensitive cells mediating the drinking response to cellular dehydration.

10.1.1. Central pathways

The osmosensitive cells which are involved in thirst lie in the lateral hypothalamus, in or near the pre-optic area (Peck and Novin 1971;

Blass and Epstein 1971) — this region is close to the supraoptic and paraventricular nuclei which are involved in ADH release. The lateral hypothalamus ('thirst centre') contains axons of several catecholaminergic neuronal systems (Ungerstedt 1971). Damage to them, produced either surgically (Teitelbaum, Cheng, and Rozin 1969) or chemically (Ungerstedt 1971; Smith, Strohmayer, and Reis 1972; Zigmond and Stricker 1972; Fibiger, Zis, and McGeer 1973), causes a period of aphagia and adipsia. The severity of the impaired drinking depends on the extent of striatal dopamine depletion caused by the lesion (Ungerstedt 1971; Zigmond and Stricker 1972), which suggests that motivated ingestive behaviour depends on the activity of dopaminergic neurons in the nigrostriatal bundle.

The ascending dopaminergic nigrostriatal bundle traverses the far lateral hypothalamic area (FLH). Injection of hypertonic saline into the pre-optic area increases the firing rate of neurons in the FLH (Morgenson and Kucharczyk 1978 and Fig. 10.1). Furthermore, lesions in the FLH (which encompass the ventromedial part of the internal capsule and the medial edge of the globus pallidus) attenuate the drinking response to both intraperitoneal and intracranial injections of hypertonic saline (Kucharczyk and Mogenson 1975; Kucharczyk, Assaf, and Mogenson 1976; Mogenson and Kucharczyk 1978). Thus it is likely that the ascending dopaminergic nigrostriatal bundle is involved in the drinking response to hydrational challenges.

The cholinergic antagonist, atropine, inhibits the drinking response to hypertonic saline (Block and Fisher 1970) whereas hexamethonium does not, and carbachol administration causes drinking (Giardina and Fisher 1971), indicating that the response to cellular dehydration is mediated, in part, by muscarinic cholinergic mechanisms.

10.2. EXTRACELLULAR FLUID DEPLETION

In addition to lowering the thirst threshold to an osmotic stimulus (see above), hypovolaemia is, itself, an effective dipsogenic stimulus. Manoeuvres such as sodium depletion (McCance 1936), haemorrhage (Fitzsimons 1961), administration of β-adrenoceptor agonists (Lehr, Mallow, and Krukowski 1967), sequestration of extracellular fluid by intraperitoneal injections of colloids (Stricker 1968), and constriction of the inferior vena cava (Fitzsimons 1964) all reduce the effective circulating blood volume, without causing cellular dehydration, and all

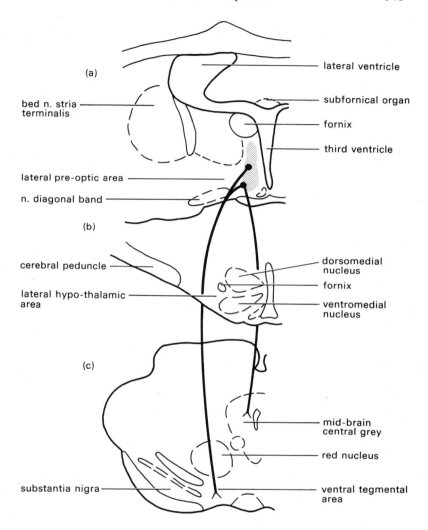

Fig. 10.1. Diagrammatic representation of the caudal part of the medial preoptic area and rostral tip of the anterior hypothalamic area (stippled region in section a) which is involved in the dipsogenic effects of angiotensin II. Two pathways arising from this area are identified. One passes caudally as the medial component of the medial forebrain bundle to terminate in the ventromedial nucleus, the mamillary body, and the ventral tegmental area. The other passes through the periventricular region and posterior hypothalamic area to end in the midbrain central grey. (With permission, Mogenson and Kucharczyk (1978).)

cause drinking. However, the drinking which occurs in response to hypovolaemia can be influenced by osmotic effects. Subcutaneous administration of a fluid with a high colloid osmotic pressure (such as polyethylene glycol) causes withdrawal of protein-free plasma into the interstitium, and hence a plasma-volume deficit (Stricker 1968). The drinking which occurs under those conditions stops before the plasma deficit is restored, unless the animals are allowed to ingest isotonic saline, in which case drinking continues until the plasma volume is normal (Stricker 1969). Presumably the ingestion of water causes a hypo-osmolality which exerts some restraint on water intake stimulated by the isotonic hypovolaemia (Stricker 1969).

10.2.1. Renin–angiotensin system

The above-mentioned stimuli which cause hypervolaemia and drinking also activate the renin–angiotensin system. Fitzsimons (1964) reported that nephrectomy markedly reduced the drinking which occurred in response to inferior vena caval ligation in rats. Administration of renal cortical extracts (Fitzsimons 1969) or angiotensin II (Fitzsimons and Simons 1969) to the nephrectomized rats restored the drinking response. Against this background it was suggested that the release of renal renin and the subsequent formation of angiotensin II contributed to the drinking response elicited by hypovolaemia (Fitzsimons 1969), despite an earlier report in which haemorrhage was shown to cause drinking in both intact and nephrectomized rats (Fitzsimons 1961). There is a direct correlation between water intake and plasma renin activity following manoeuvres which reduce plasma volume (Leenen and Stricker 1974). Meyer, Rauscher, Peskar, and Hertting (1973) showed that isoprenaline caused drinking and a concurrent increase in renin release in rats; both these responses were blocked by propranolol. Furthermore, intramuscular phentolamine caused drinking and renin release and these responses were suppressed by ganglion blockade (Meyer *et al.* 1973) and enhanced by inhibition of noradrenalin re-uptake (Meyer and Hertting 1975). These workers suggested that β-adrenoceptor stimulation (either directly or through reflex sympathetic activation due to the hypotensive effect of phentolamine) was responsible for activating the renin–angiotensin system which subquently triggered a drinking response (Meyer *et al.* 1973; Meyer and Hertting 1975 and Fig. 10.2).

Further support for the role of the renin–angiotensin system in

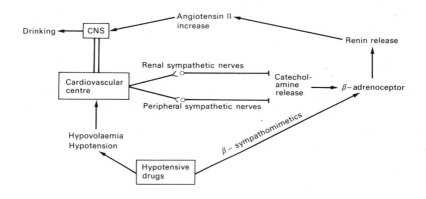

Fig. 10.2. The β-sympathomimetic isoprenaline acts directly on the β-receptors which trigger renin release. Other dipsogens such as phentolamine and hydralazine cause a reflexly mediated catecholamine-release from the sympathetic nervous system. The stimuli for this release may be hypovolaemia or hypotension. These catecholamines act on the β-receptors which are responsible for renin release. The increase in plasma renin concentration induced either by β-mimetics or by catecholamine-releasing agents causes an enhanced generation of angiotensin I and II. These peptides mediate drinking directly or indirectly. (With permission, Meyer and Hertting (1975).)

drinking was the demonstration that intravenous (Fitzsimons and Simons 1969) or central (Epstein, Fitzsimons, and Rolls 1970) administration of valine-5-angiotensin to rats caused drinking. It was claimed that the doses used in those studies were far outside the physiological range (Stricker, Bradshaw, and McDonald 1976), but Hsiao, Epstein, and Camardo (1977) have since shown that intravenous infusions of physiological doses of the naturally occurring angiotensin (aspartic acid-1, isoleucine-5-) cause drinking in rats.

The drinking response to angiotensin I is greater than that to angiotensin II (Epstein 1972), probably because of the longer half-life of the former (Osborne, Pooters, Angles d'Auriac, Epstein, Worcel, and Meyer 1971). However, the dipsogenic effect of angiotensin I depends on its conversion to angiotensin II in the brain (Lehr, Goldman, and Casner, 1973; Epstein, Fitzsimons, and Johnson 1974).

The work outlined above was performed in rats. It was previously thought that angiotensin II was not an effective dipsogen in the dog (Kozlowski et al. 1972; Szczepańska-Sadowska and Fitzsimons 1975) but in those studies, the infusion rates were very low. Fitzsimons, Kucharczyk, and Richards (1978) have since shown that drinking occurs in dogs when angiotensin II is infused at a rate high enough to produce physiological concentrations in the circulation. This is consistent with the work of Rolls and Ramsay (1975) in which dogs with renal hypertension (and elevated plasma renin levels) had increased fluid intakes; those workers suggested that angiotensin II was involved in the drinking response.

Fitzsimons et al. (1978) found that systemic infusions of angiotensin II were more effective dipsogenic stimuli than infusions of renin or angiotensin I in dogs (contrary to the findings in the rat — see above). Furthermore, intravenous infusions of angiotensin I were more potent than intracarotid infusions, and it was concluded that angiotensin I must be converted to the octapeptide in the periphery before reaching the receptor sites (Fitzsimons et al. 1978). Since central administration of angiotensin II causes drinking in dogs (Fitzsimons and Kucharzcyk 1978) it is likely that these receptors are located in the brain.

10.2.2. Central pathways

Three central sites have been reported to mediate the dipsogenic effects of angiotensin: the pre-optic area (Epstein et al. 1970; Kucharczyk and Mogenson 1976); the sub-fornical organ (Simpson and Routtenberg 1973; Simpson 1975; Mangiapane and Simpson 1977; Simpson, Mangiapane, and Dellmann 1978); and the organum vasculosum lamina terminalis (OVLT; Buggy and Johnson 1977 and Fig. 10.3).

The original evidence for the involvement of the pre-optic area came from work in which angiotensin II injected into that region was shown to cause drinking (Epstein et al. 1970). However, Simpson (1975) argued that, in previous studies, cannulae implanted into the pre-optic area always passed through the cerebral ventricles and angiotensin II may have refluxed up the outside of the cannula shaft into the ventricles; the peptide could then reach some other site through the cerebrospinal fluid. In support of this, Johnson (1975) showed that injection of angiotensin II into the pre-optic area, using cannulae which bypassed the ventricles, did not cause drinking. But Kucharczyk and Mogenson (1976) applied angiotesin II to the pre-optic area with

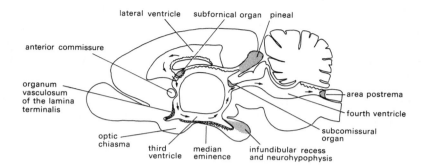

Fig. 10.3. The circumventricular organs (shaded) shown in a diagrammatic saggital section of an adult rat brain. Arrows indicate the direction of flow of cerebrospinal fluid. (With permission, Phillips (1978).)

cannulae angled such that they did not cross the cerebral ventricles and still obtained a drinking response. Moreover, Fitzsimons and Kucharczyk (1978) showed, in the dog, that injection of angiotensin II into the pre-optic area caused drinking at a much lower dose than when the injection was made directly into the ventricles. They argued that an injection into the pre-optic area could not have refluxed along the cannula shaft, entered the ventricles and caused a greater drinking response than did an injection directly into the ventricles (Fitzsimons and Kucharzcyk 1978).

It is doubtful that circulating angiotensin could reach the pre-optic area, but there is strong evidence that the caudate nucleus (lateral to the pre-optic area) is a region where angiotensin is produced (Ganten, Marquez-Julio, Granger, Hayduk, Karsunky, Boucher, and Genest 1971; Fischer-Ferraro, Nahmod, Goldstein, and Finkielman 1971); this angiotensin might act on the pre-optic area to cause drinking (Kucharczyk et al. 1976).

Drinking induced by injection of angiotensin II into the pre-optic area is attenuated by administration of the dopamine antagonist, haloperidol (Fitzsimons and Setler 1971), and by lesions placed in the medial aspect of the lateral hypothalamus (MLH; Fig. 10.4; Kucharczyk and Mogenson 1975; Kucharczyk et al. 1976; Mogenson and Kucharczyk 1978). Furthermore, there is an increased firing rate in neurons situated in the MLH when angiotensin II is injected into the pre-optic area (Mogenson and Kucharczyk 1978). Using autoradiographic techniques,

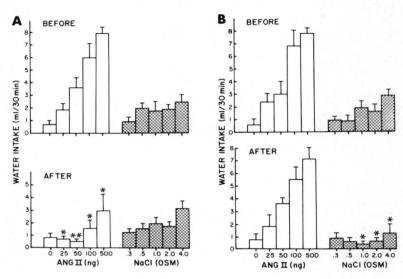

Fig. 10.4. Differential effects of lesions of the midlateral hypothalamus and far-lateral hypothalamus on water intake after microinjections of angiotensin II and hypertonic saline into the preoptic region through cannulae positioned so as to bypass the cerebral ventricles. (A) Water intake (ml/30 min) of 17 rats before (top panel) and after (lower panel) lesions of the midlateral hypothalamus. (B) Water intake of 13 rats before (top panel) and after (lower panel) lesions of the far-lateral hypothalamus. Results are given as means ± SE. *$p<0.05$ compared to prelesion intakes **$p<0.01$ compared to prelesion intakes. (With permission, Mogenson and Kucharczyk (1978).)

Mogenson and Kucharczyk (1978) demonstrated two distinct neural pathways involved in the drinking response to angiotensin II administered to the pre-optic area. One pathway passes caudally as the medial component of the medial forebrain bundle and terminates in the MLH, the ventromedial nucleus, the mammillary body and the ventral tegmental area (Fig. 10.1). The other descends through the periventricular region and posterior hypothalamic area to end in the midbrain central grey.

The OVLT may also mediate the effects of central angiotensin II. Intraventricularly administered angiotensin II does not cause drinking when the OVLT is occluded (Hoffman and Phillips 1976; Buggy, Fink, Johnson, and Brody 1977), whereas drinking occurs with as little as 50 fg of angiotensin II injected into the OVLT (Phillips 1978).

Ablation of the subfornical organ reduces the drinking response to intravenous angiotensin II (Abdelaal, Assaf, Kucharczyk, and Mogenson 1974) and to extracellular fluid depletion (Simpson *et al.* 1978). However, when angiotensin II is injected into the cerebral ventricles, at a site close to the ependymal surface of the subfornical organ, there is no significant drinking response (Hoffman and Phillips 1976). The subfornical organ is an area which lies outside the blood-brain barrier and it is likely that it is responsible for mediating the dipsogenic effects of circulating angiotensin II. Centrally produced angiotensin II may not have ready access to this region (Phillips 1978). Drinking caused by injection of angiotensin II into the subfornical organ is not affected by lesions in the MLH (Kucharczyk *et al.* 1976) and therefore cannot involve the same neural pathways as the drinking response to angiotensin II administered to the pre-optic area. There is evidence of neural projections from the subfornical organ to the OVLT and supraoptic nucleus (Shapiro and Miselis 1978) but the identity of the transmitter substances and the physiological importance of these neural links is, as yet, speculative (Simpson *et al.* 1978).

It is generally agreed that drinking induced by angiotensin II does not depend on cholinergic neurons (Giardina and Fisher 1971; Fitzsimons and Setler 1971) although Severs, Summy-Long, Taylor, and Connor (1970) reported that atropine blocked the dipsogenic effects of intraventricular angiotensin II; the reason for this discrepancy is unclear. Intraventricular administration of phentolamine (Severs, Summy-Long, Daniels-Severs, and Connor 1971) or 6-hydroxydopamine (Fitzsimons 1975; Stricker and Zigmond 1975) attenuates the drinking response to angiotensin II administered by that route. However, intraventricular 6-hydroxydopamine does not affect the drinking response to subcutaneous isoprenaline or polyethylene glycol (Sticker and Zigmond 1975). Thus it is likely that different receptor sites and different neural pathways are involved in the dipsogenic effects of peripheral and central angiotensin.

REFERENCES

Abdelaal, A. E., Assaf, S. Y., Kucharczyk, J., and Mogenson, G. J. (1974). Effect of ablation of the subfornical organ on water intake elicited by systemically administered angiotensin II. *Can. J. Physiol. Pharmacol.* **52**, 1217–20.

Andersson, B. (1978). Sodium versus osmotic sensitivity in cerebral control of water balance. In *Osmotic and volume regulation. Proc.*

Alfred Benzon Symposium XI (ed. C. B. Jorgensen and E. Skadhauge), pp. 84–94. Academic Press, London.
——, Dallman, M. F., and Olsson, K. (1969). Observations on central control of drinking and of the release of antidiuretic hormone (ADH). *Life Sci.* **8**, 425–32.
Blass, E. M. and Epstein, A. N. (1971). A lateral preoptic osmosensitive zone for thirst in the rat. *J. comp. physiol. Psychol.* **76**, 378–94.
Block, M. L. and Fisher, A. E. (1970). Anticholinergic central blockade of salt-aroused and deprivation-induced drinking. *Physiol. Behav.* **5**, 525–7.
Buggy, J. and Johnson, A. K. (1977). Preoptic-hypothalamic periventricular lesions: thirst deficits and hypernatremia. *Amer. J. Physiol.* **233**, R44–R52.
——, Fink, G. D., Johnson, A. K., and Brody, M. J. (1977). Prevention of the development of renal hypertension by anteroventral third ventricular tissue lesions. *Circulation Res.* **40**, (Suppl. I), 110–17.
Epstein, A. N. (1972). Drinking induced by low doses of intravenous angiotensin. *The Physiologist* **15**, 127.
——, Fitzsimons, J. T., and Johnson, A. K. (1974). Peptide antagonists of the renin–angiotensin system and the elucidation of the receptors for angiotensin-induced drinking. *J. Physiol.* **238**, 34P–35P.
——, ——, and Rolls, B. J. (1970). Drinking induced by injection of angiotensin into the brain of the rat. *J. Physiol.* **210**, 457–74.
Fibiger, H. C., Zis, A. P., and McGeer, E. G. (1973). Feeding and drinking deficits after 6-hydroxydopamine administration in the rat: similarities to the lateral hypothalamic syndrome. *Brain Res.* **55**, 135–48.
Fischer-Ferraro, C., Nahmod, V. E., Goldstein, D. J., and Finkielman, S. (1971). Angiotensin and renin in rat and dog brain. *J. exp. Med.* **133**, 353–61.
Fitzsimons, J. T. (1961). Drinking by rats depleted of body fluid without increase in osmotic pressure. *J. Physiol.* **159**, 297–309.
—— (1964). Drinking caused by constriction of the inferior vena cava in the rat. *Nature (Lond.)* **204**, 479–80.
—— (1969). The role of a renal thirst factor in drinking induced by extracellular stimuli. *J. Physiol.* **201**, 349–68.
—— (1975). The renin–angiotensin system and drinking behaviour. *Prog. Brain Res.* **42**, 215–33.
—— and Kucharczyk, J. (1978). Drinking and haemodynamic changes induced in the dog by intracranial injection of components of the renin–angiotensin system. *J. Physiol.* **276**, 419–34.
——, —— and Richards, G. (1978). Systemic angiotensin-induced drinking in the dog: a physiological phenomenon. *J. Physiol.* **276**, 435–48.

Fitzsimons, J. T. and Setler, P. E. (1971). Catecholaminergic mechanisms in angiotensin-induced drinking. *J. Physiol.* **218**, 43P–44P.
—— and Simons, B. J. (1969). The effect on drinking in the rat of intravenous infusion of angiotensin, given alone or in combination with other stimuli of thirst. *J. Physiol.* **203**, 45–57.
Ganten, D., Marquez-Julio, A., Granger, P., Hayduk, K., Karsunky, K. P., Boucher, R., and Genest, J. (1971). Renin in dog brain. *Amer. J. Physiol.* **221**, 1733–7.
Giardina, A. R. and Fisher, A. E. (1971). Effect of atropine on drinking induced by carbachol, angiotensin and isoproterenol. *Physiol. Behav.* **7**, 653–5.
Gilman, A. (1937). The relation between blood osmotic pressure, fluid distribution, and voluntary water intake. *Amer. J. Physiol.* **120**, 323–8.
Hoffman, W. E. and Phillips, M. I. (1976). The effect of subfornical organ lesions and ventricular blockade on drinking induced by angiotensin II. *Brain Res.* **108**, 59–73.
Hsiao, S., Epstein, A. N., and Camardo, J. S. (1977). The dipsogenic potency of peripheral angiotensin II. *Hormones Behav.* **8**, 129–40.
Johnson, A. K. (1975). The role of the cerebral ventricular system in angiotensin-induced thirst. In *Control mechanisms in drinking* (ed. G. Peters, J. T. Fitzsimons, and L. Peters-Haefeli), pp. 117–22, Springer Verlag, Heidelberg.
Kozlowski, S., Drzewiecki, K., and Zurawski, W. (1972). Relationship between osmotic reactivity of the thirst mechanism and the angiotensin and aldosterone level in the blood of dogs. *Acta physiol. pol.* **23**, 369–76.
—— and Szczepańska-Sadowska, E. (1975). Mechanisms of hypovolaemic thirst and interactions between hypovolaemia, hyperosmolality and the antidiuretic system. In *Control mechanisms in drinking* (ed. G. Peters, J. T. Fitzsimons, and L. Peters-Haefeli), pp. 25–35, Springer-Verlag, Heidelberg.
Kucharczyk, J. and Mogenson, G. J. (1975). Separate lateral hypothalamic pathways for extracellular and intracellular thirst. *Amer. J. Physiol.* **228**, 295–301.
——, —— (1976). Specific deficits in regulatory drinking following electrolytic lesions of the lateral hypothalamus. *Exp. Neurol.* **53**, 371–85.
——, Assaf, S. Y., and Mogenson, G. J. (1976). Differential effects of brain lesions on thirst induced by the administration of angiotensin-II to the preoptic region, subfornical organ and anterior third ventricle. *Brain Res.* **108**, 327–37.
Leenen, F. H. and Stricker, E. M. (1974). Plasma renin activity and thirst following hypovolemia or caval ligation in rats. *Amer. J. Physiol.* **226**, 1238–42.
Lehr, D., Goldman, H. W., and Casner, P. (1973). Renin–angiotensin

role in thirst: paradoxical enhancement of drinking by angiotensin converting enzyme inhibitor. *Science* **182**, 1031–4.

Lehr, D., Mallow, J., and Krukowski, M. (1967). Copious drinking and simultaneous inhibition of urine flow elicited by beta-adrenergic stimulation and contrary effect of alpha-adrenergic stimulation. *J. Pharmac. exp. Ther.* **158**, 150–63.

Mangiapane, M. L. and Simpson, J. B. (1977). Subfornical organ: site of drinking and pressor effects of angiotensin II. *Neurosci. Abst.* **3**, 351.

McCance, R. A. (1936). Experimental sodium chloride deficiency in man. *Proc. roy. Soc.* **119**, 245–68.

Meyer, D. K. and Hertting, G. (1975). Drinking induced by direct or indirect stimulation of beta-receptors: evidence for involvement of the renin–angiotensin system. In *Control mechanisms in drinking* (ed G. Peters, J. T. Fitzsimons, and L. Peters-Haefeli), pp. 89–93, Springer-Verlag, Heidelberg.

——, Rauscher, W., Peskar, B., and Hertting, G. (1973). The mechanism of the drinking response to some hypotensive drugs: activation of the renin–angiotensin system by direct or reflex-mediated stimulation of β-receptors. *Naunyn Schmiedebergs Arch. Pharmacol.* **276**, 13–24.

Mogenson, G. J. and Kucharczyk, J. (1978). Central neural pathways for angiotensin-induced thirst. *Fed. Proc.* **37**, 2683–8.

Olsson, K. (1972). Dipsogenic effects of intracarotid infusions of various hyperosmolal solutions. *Acta physiol. Scand.* **85**, 517–22.

Osborne, M. J., Pooters, N., Angeles d'Auriac, G., Epstein, A. N., Worcel, M., and Meyer, P. (1971). Metabolism of tritiated angiotensin II in anaesthetized rats. *Pflügers' Arch.* **326**, 101–14.

Peck, J. W. and Novin, D. (1971). Evidence that osmoreceptors mediating drinking in rabbits are in the lateral preoptic area. *J. comp. physiol. Psychol.* **74**, 134–47.

Phillips, M. I. (1978). Angiotensin in the brain. *Neuroendocrinol.* **25**, 354–77.

Rolls, B. J. and Ramsay, D. J. (1975). The elevation of endogenous angiotensin and thirst in the dog. In *Control mechanisms in drinking* (ed. G. Peters, J. T. Fitzsimons, and L. Peters-Haefeli), pp. 74–8. Springer-Verlag, Heidelberg.

Severs, W. B., Summy-Long, J., Daniels-Severs, A., and Connor, J. D. (1971). Influence of adrenergic blocking drugs on central angiotensin effects. *Pharmacol.* **5**, 205–14.

——, ——, Taylor, J. S., and Connor, J. D. (1970). A central effect of angiotensin: release of pituitary pressor material. *J. Pharmacol. exp. Ther.* **174**, 27–33.

Shapiro, R. E. and Miselis, R. R. (1978). Confirmation of the subfornical organ projection to the supraoptic nucleus in the rat. *Anat. Rec.* **190**, 538–9.

Simpson, J. B. (1975). Subfornical organ involvement in angiotensin-induced drinking. In *Control mechanisms of drinking* (ed. G. Peters, J. T. Fitzsimons, and L. Peters-Haefeli), pp. 123–6. Springer-Verlag, Heidelberg.

—— and Routtenberg, A. (1973). Subfornical organ: site of drinking elicitation by angiotensin II. *Science* **181**, 1172–5.

——, Mangiapane, M. L., and Dellmann, H.-D. (1978). Central receptor sites for angiotensin-induced drinking: a critical review. *Fed. Proc.* **37**, 2676–82.

Smith, G. P., Strohmayer, A. J., and Reis, D. J. (1972). Effect of lateral hypothalamic injections of 6-hydroxydopamine on food and water intake in rats. *Nature (Lond.)* **235**, 27–9.

Stricker, E. M. (1968). Some physiological and motivational properties of the hypovolemic stimulus for thirst. *Physiol. Behav.* **3**, 379–85.

—— (1969). Osmoregulation and volume regulation in rats: inhibition of hypovolemic thirst by water. *Amer. J. Physiol.* **217**, 98–105.

——, Bradshaw, W. G., and McDonald, R. H., Jr. (1976). The renin-angiotensin system and thirst. A re-evaluation. *Science* **194**, 1169–71.

—— and Zigmond, M. J. (1975). Brain catecholamines and thirst. In *Control mechanisms in drinking* (ed. G. Peters, J. T. Fitzsimons, and L. Peters-Haefeli), pp. 55–61. Springer-Verlag, Heidelberg.

Szczepańska-Sadowska, E. and Fitzsimons, J. T. (1975). The effects of angiotensin II, renin and isoprenaline on drinking in the dog. In *Control mechanisms in drinking* (ed. G. Peters, J. T. Fitzsimons, and L. Peters-Haefeli), pp. 69–73. Springer-Verlag, Heidelberg.

——, Kozlowski, S., and Sobacińska, J. (1974). Blood antidiuretic hormone level and osmotic reactivity of the thirst mechanism in dogs. *Amer. J. Physiol.* **227**, 766–70.

Teitelbaum, P., Cheng, M.-F., and Rozin, P. (1969). Stages of recovery and development of lateral hypothalamic control of food and water intake. *Ann. NY Acad. Sci.* **157**, 849–58.

Thames, M. D. and Schmid, P. G. (1979). Cardiopulmonary receptors with vagal afferents tonically inhibit ADH release in the dog. *Amer. J. Physiol.* **237**, H229–H304.

Ungerstedt, U. (1971). Adipsia and aphagia after 6-hydroxydopamine induced degeneration of the nigro-striatal dopamine system. *Acta physiol. Scand.* Suppl. **367**, 95–122.

Zigmond, M. J. and Stricker, E. M. (1972). Deficit in feeding behaviour after intraventricular injection of 6-hydroxydopamine in rats. *Science* **177**, 1211–14.

Index

active chloride transport 132, 137
active sodium transport 132
adrenalectomy
   and administration of aldosterone 81, 84
   effect on renin release 92
adrenal cortex,
   effect of potassium balance on 87
   effect of sodium restriction on 85
   secretion of aldosterone by 81
adrenal catecholamines
   and the hypothalamic pressor area 25
   and the medullary cardiovascular area 21
   effect of angiotensin II on the release of 105, 109, 110
adrenalin
   as a transmitter in the medullary cardiovascular area 22
   effect on hypothalamic depressor area 28-9
   effect on renin release 98
adrenocorticotrophic hormone, effect on aldosterone secretion 84, 86-7
afferent arteriole,
   renin release from 92, 95-6
   role in renal autoregulation 53, 59-60
   vasoconstriction of 93
   vasodilatation of 53, 60
aldosterone 81-6
   effect of angiotensin II on secretion of 74
   mechanism of action
      metabolic theory 82
      permease theory 82
      pump theory 82
   receptor complex 82
amygdala 25
angiotensin I
   converting enzyme 105
   converting enzyme inhibitor 85, 106
   effects on thirst 143-4
   formation of 57, 105
angiotensin II
   and central nervous system 108-9
   antagonist to 85, 106, 110
   antibodies to 107
   effects on aldosterone release 85-7
   effects on antidiuretic hormone release 123, 126
   effects on baroreflex responses 110
   effects on renin release 99
   effects on thirst 139, 142-7
   formation of 105
   in glomerulo-tubular balance 73
   in renal autoregulation 57-60
   vasoconstrictor effects of 105, 107
angiotensin III, effects on aldosterone secretion 85
antidiuretic hormone,
   adrenergic mechanisms and release of 118
   and angiotensin II 74, 109
   cholinergic mechanisms and release of 118
   effect of antimitotic agents on action of 125
   effect of volume depletion on release of 139
   effect on renin release 99
   effect on thirst threshold 139
   effect on urine concentration 134-6
   synthesis of 114
aortic
   arch baroreceptors 17, 30-2
      B-fibres of 34-5
      C-fibres of 34-5
   body 44
   nerve 21, 29, 31-2, 35
   transmural pressure 17
area postrema 108, 109
arterial baroreceptors 17-21, 29, 31, 35-6, 42-4
   and antidiuretic hormone release 119, 121
   role in capillary pressure regulation 7
   role in glomerulo-tubular balance 75
atrial receptors 37-41
   and antidiuretic hormone release 119, 121

Index

and renin release 98
and thirst 139
with myelinated nerve fibres 37–8, 41–3
with unmyelinated nerve fibres 37–8, 41–3

baroreceptor denervation
  and antidiuretic hormone release 121
  causing hypertension 32
baroreceptors (*see also* arterial baroreceptors)
  central connections of 20
  defence reaction and 25
  effects of extracellular sodium on 19, 107
  effects of pulse frequency on 17
  effects of pulse pressure on 17
Bezold–Jarisch effect 42
bicarbonate,
  effects on sodium reabsorption 67
  proximal tubular reabsorption of 64, 66
blood–brain barrier, glucose transport across 116
blood flow,
  metabolic autoregulation of 5–6
  myogenic autoregulation of 4–5
blood volume
  depletion,
    effects on antidiuretic hormone release 119, 121
    effects on thirst 139
  expansion and secretion of natriuretic hormone 87
Bowman's capsule, hydrostatic pressure of 52
bulbospinal neurons 22

calcium
  and antidiuretic hormone release 115
  and proximal tubular sodium reabsorption 72
  and renin release 100
capillary
  colloid osmotic pressure and countercurrent exchange 136
  filtration coefficient 1, 2, 8
  hydrostatic pressure 1, 3, 6–9, 11
    and countercurrent exchange 135–6

cardiac mechanoreceptors 37–42
  role in capillary pressure regulation 7
carotid
  body 44
  sinus 17, 21, 26, 30–2
    effects of alpha-adrenoceptor agonists on 19
    nerve, activity in 23, 29, 30–2, 36
    pressure and baroreceptor activity 19
    pressure and splanchnic venous capacity 36
    sodium sensitivity of 107
caudate nucleus 145
chemoreceptors 43–4
chloride,
  diffusion potential of 65
  proximal tubular reabsorption of 64, 67
cholinergic
  nerves and thirst 140
  vasodilatation and arterial baroreceptors 30
  vasodilatation and chemoreceptors 44
  vasodilatation in the defence reaction 26
cingulate gyrus 25
circumventricular organs 108
collecting duct, renal,
  action of antidiuretic hormone on 124
  urine concentration in 132
cyclic AMP and action of antidiuretic hormone 124

defence reaction 25–6
deoxycorticosterone
  and hypertension, noradrenalin storage in 107
  and hypertension, role of angiotensin 110
  and hypertension, role of antidiuretic hormone 126
  escape from effects of 84
  renal autoregulation after treatment with 59
descending noradrenergic fibres 22
diabetes insipidus 121
distal
  tubular load and renin release 94–7

tubule,
    effects of aldosterone on 81–4
    sodium reabsorption in 83
dorsal motor nucleus 21, 108
dorsolateral funiculus 22

efferent arteriole
    and juxtaglomerular apparatus 55
    renin release from 57, 94
    role in glomerulo-tubular balance 73
    role in renal autoregulation 59, 60
extracellular fluid 75, 114
    osmolality and antidiuretic hormone release 114
    osmolality and capillary fluid flux 9
    sodium concentration in 52
    volume and aldosterone release 85
    volume and angiotensin II levels 107
    volume and renin release 96
    volume depletion and thirst 140, 147
    volume expansion 84, 96

filtration fraction 71
free water clearance 114

glomerular filtration rate 52–60, 71–5
    and countercurrent exchange 137
    and renin release 92, 97
glomerulo-tubular balance
    in the distal tubule 80
    in the proximal tubule 63, 70–5
glossopharyngeal nerve 20, 27

haemorrhage
    and adrenal medullary catecholamine release 109–10
    and aldosterone release 85
    and antidiuretic hormone release 119, 121, 123, 126
    and renin release 98
    atrial receptor signalling in 38
    capillary fluid flux in 8
    chemoreceptor activation in 44–5
    hyperglycaemia following 9
    protein synthesis following 11
    proximal tubular sodium reabsorption in 74–5
hepatic glycogenolysis 9
6-hydroxydopamine 23

5-hydroxytryptamine 23
hypertension,
    aldosterone levels in 85
    angiotensin II levels in 110, 144
    antidiuretic hormone levels in 121, 126
    baroreceptor denervation and 32
    deoxycorticosterone and 107–8
    NTS lesions and 23
    renal 84
hypertonic
    saline,
        effect on antidiuretic hormone release 115, 123
        effect on thirst 139–40
    sucrose, effect on antidiuretic hormone release 115
    urea,
        effect on antidiuretic hormone release 115
        effect on thirst 139
hypothalamus,
    alpha- and beta-adrenoceptors in 28
    cardiovascular control by 23–9
    control of thirst by 139, 140, 145–7
    depressor area of 23–4

intermedio-lateral cell column 22
interstitial
    fluid 9, 80
        colloid osmotic pressure in 1, 10, 68
        hydrostatic pressure in 1, 9, 68
        osmolality and countercurrent exchange 135–6
        osmolality and urine concentration 132, 134
    pressure,
        effects on glomerulo-tubular balance 72
        effects on renal autoregulation 53
intratubular environment,
    effects on glomerulo-tubular balance 72–3
    effects on proximal tubular reabsorption 68
inulin 70
isoprenaline
    causing renin release 98–9
    causing thirst 147

effects on capillary pressure
regulation 7

juxtaglomerular
  apparatus 55
    sympathetic innervation of 98
  cells,
    location of 84
    renin release from 92, 96–100
juxtamedullary nephrons, effect of antidiuretic hormone on blood flow in 136

lateral funiculus 22
lateral-intercellular spaces 67–9
loop of Henle 80
  permeability of descending limb of 131
  permeability of thick ascending limb of 132
  permeability of thin ascending limb of 131
  thick ascending limb of 80
  thin ascending limb of 80

macula densa
  and glomerulo-tubular balance 73
  and renal autoregulation 55–6
  and renin release 58, 92–4
medial forebrain bundle 146
medulla,
  cardiovascular control by the 21–9
  depressor areas of the 21–2
  pressor areas of the 21
microtubules, microfilaments, and action of antidiuretic hormone 125

natriuresis following extracellular fluid volume expansion 69, 75, 84
natriuretic hormone 84, 87
neuropeptides and cardiovascular control 29
neurophysins 115
nigrostriatal bundle 140
nodose ganglion 21
non-electrolyte solutions
  and proximal tubular sodium reabsorption 63
  and renal autoregulation 57
noradrenalin,
  action on hypothalamic cardio-vascular centres 28–9

  action on medullary cardiovascular centres 23
  and supraoptic neuron activity 118
  angiotensin II causing release of 105
  diuresis following 121
  effects on arterial baroreceptors 19
  effects on renin release 98
  storage and sodium intake 107–8
nucleus
  ambiguus 21, 26
  of the tractus solitarius
    area postrema connections with 108
    baroreceptor inputs to 20, 26

orbital cortex 25
organum vasculosum lamina terminalis 144, 146
osmolality and renin release 97
osmoreceptors,
  role in antidiuretic hormone release 115–16
  role in thirst 139

parathyroid hormone 72
paraventricular nucleus
  and release of antidiuretic hormone 117, 119
  and synthesis of antidiuretic hormone 114–15
peritubular capillaries
  and extracellular fluid volume expansion 84
  influence on proximal tubular reabsorption 68–9
  involvement in glomerulo-tubular balance 71
plasma
  colloid osmotic pressure 1, 10–11
    effects on glomerular filtration 52
    effects on proximal tubular reabsorption 69, 72
  osmolality,
    effects on antidiuretic hormone release 118, 121–3
    effects on thirst 139
posterior pituitary, antidiuretic hormone storage in 114
post-glomerular resistance and counter-current exchange 136–7

Index

potassium,
  aldosterone and excretion of 81, 83
  and secretion of aldosterone 84, 86–7
  effects on renin release 97
pre- to post-capillary resistance ratio 7–8
pre-optic area and thirst 139, 140, 144, 146–7
prostaglandins,
  effects on aldosterone release 86
  effects on antidiuretic hormone release 123–4
  effects on intrarenal blood flow 136
  effects on renin release 99
protein synthesis inhibitors and action of aldosterone 82–4
proximal
  tubule 63–4
  tubular reabsorption 68–75, 97
    and extracellular fluid volume expansion 84
    and osmotic diuresis 63
    effects of aldosterone on 82
    effects of alpha-adrenoceptor antagonist on 74–5
    effects of ouabain on 63
    osmotic gradient and 66–7
pulmonary
  receptors 43
  vein-atrial junction 38

Raphe nucleus 22
renal
  arterial pressure
    and renal autoregulation 60, 95–6
    and renin release 58, 95–6
  autoregulation 52, 54, 58
    hormonal mechanisms and 54
    myogenic mechanisms and 53, 60
  blood flow 96
    autoregulation of 96
    distribution of 136
  countercurrent
    exchange 135–7
    multiplication 132–5
  nerve activity,
    effects of arterial baroreceptors on 34–5
    effects of atrial receptors on 38
    effects of ventricular receptors on 42
  oxygen consumption 63
  plasma flow,
    hormonal autoregulation of 54–60
    myogenic autoregulation of 52–3
  vasodilatation and atrial receptors 38
renin 57, 73
  activity in the central nervous system 108
  effects of sodium restriction on 105, 107
  involvement in thirst 142
  release,
    baroreceptor hypothesis of 92, 96–7
    effect of body fluid balance on 97
    effect of catecholamines on 98
    effect of potassium on 87
    excitation–contraction coupling and 100
    extracellular fluid volume expansion and 85
    macula densa hypothesis of 94, 96–7
    mechanism of 100
    perfusion pressure and 92–5, 99
    right atrial pressure and 98
    sodium depletion and 97
    stimulus–secretion coupling and 100
    sympathetic nerves and 99

sodium,
  active transport and reabsorption of 63–6, 69
  and aldosterone release 85–7
  and angiotensin II release 106–8
  and renal autoregulation 56–8, 60
  and renin release 94, 96–7
  and thirst 140
  balance and aldosterone 85–7
  baroreflex control of excretion of 19
  cerebrospinal fluid concentration of 116

extracellular fluid concentration
 of 52
glomerulotubular balance and
 excretion of 73, 75
sensor,
 role in antidiuretic hormone
  release 116, 123
 role in thirst 139
 sympathetic nerves and reabsorption
  of 74-5
subfornical organ 144, 146
subnucleus medialis 109
supraoptic nucleus
 and antidiuretic hormone 114,
  117, 119
 and thirst 147
 effects of arterial baroreceptors on
  121
sympathetic
 afferents 42-3
 nerve activity,
  effects of angiotensin II on
   108-9
  effects of atrial receptors on 38
  effects on proximal tubular
   reabsorption 74-5
  medullary cardiovascular centres
   and 21
 pre-ganglionic neurons 22, 26
systemic arterial blood pressure,
 arterial baroreceptors and control
  of 32
 central effects of angiotensin II on
  108, 110
 defence reaction and 25
 effects of antidiuretic hormone on
  126
 peripheral effects of angiotensin II
  on 86, 97, 105, 107, 110

thirst,
 beta-adrenoceptor stimulation and
  142
 cellular dehydration and 139
 centre 139
 effects of angiotensin I on 143
 effects of angiotensin II on 146-7
 effects of atropine on 147
 effects of 6-hydroxydopamine on
  147
 effects of isoprenaline on 147
 effects of phentolamine on 147

extracellular fluid volume depletion
 and 147
hypovolaemia and 142
muscarinic receptors and 140
role of osmoreceptors in 139
role of renin–angiotensin system in
 142
threshold 139-40
transcapillary fluid movement 3, 8,
 11, 135

urea permeability,
 effects of antidiuretic hormone on
  124-5, 136
 of the collecting duct 134
 of the loop of Henle 131
urine concentration,
 effects of antidiuretic hormone on
  118, 124
 effects of renal tissue osmolality
  on 131

vagal
 afferents
  and antidiuretic hormone
   release 121
  atrial receptors with 37-8
  ventricular receptors with 41-2
 bradycardia,
  arterial baroreceptors causing
   30
  defence reaction inhibiting 26
  ventricular receptors causing 42
 efferent tone 21
vasa recta 135-7
vascular resistance 17
 effect of sympathetic constrictor
  nerves on 25, 42, 44
 effect of sympathetic dilator nerves
  on 26, 44
 metabolic autoregulation affecting
  5-6
veins 8
 and baroreflex responses 35-6
ventricular receptors 41-3
 Bezold–Jarisch effect caused by 42
 myocardial infarction and
  activation of 42
 myocardial ischaemia and
  activation of 42
volume receptors, role in antidiuretic
 hormone release 119